3/11

"*UFOs: Myths, Conspiracies, and Realities* changes the playing field for both true believers and skeptics alike. Alexander notes that disclosure has already happened, and what is really wanted is conformation. He strongly warns, be careful what you wish for when asking for presidential intervention. Success could set the field of ufology back decades."
—George Noory, host of *Coast to Coast AM*

"Dr. John B. Alexander delivers a captivating and one-of-a-kind exposé of UFO phenomena and conspiracy theories. Told from the unmatched perspective of a government insider, and combining encyclopedic knowledge with a passion for the subject matter, he takes on the long-standing debate of whether UFOs exist, while opening the door to a whole new set of mysteries. Whether you're a believer or not, this fact-based analysis is a must-read."
—Gale Anne Hurd, producer of the movies *Terminator*, *Aliens*, *The Abyes*, and *Armageddon*

"John B. Alexander's latest book, *UFOs: Myths, Conspiracies, and Realities*, is a must-read! Of the more than 200 books in the UFO bibliography, this one has to rank at the top. Finally, someone has brought this strange enigma to the front burner!"
—Col. William Coleman, Jr. (ret.), former Chief of Public Information for the U.S. Air Force and chief spokesperson for Project Blue Book

"John Alexander is the real deal. He doesn't just talk the talk—he's walked the walk. Here's a man with a top secret security clearance who researched the UFO phenomenon and discussed it at the highest levels, in the shadowy world of military and intelligence which he's inhabited for decades. Forget everything you think you know about UFOs—this insider's account exposes the reality. And it's a reality that will come as a big surprise. Packed with top-grade information, insightful analysis, and fascinating anecdotes, Alexander's interesting and controversial book sets the gold standard for titles on this subject."
—Nick Pope, author of *Open Skies, Closed Minds*

"In these pages, Dr. Alexander describes the ultimate exploration of the government's UFO secrets: that of an insider with the clearances and credentials to discover the truth. And the truth that he has discovered is deeply shocking: somebody—or something—is certainly here, but we have no ongoing response to this mysterious presence at all. Irrefutable cases confirm a UFO reality, but official response has been sketchy at best, and no effort has been made at all to determine whether this reality represents a threat to the people of Earth. For anybody interested in the UFO mystery, or concerned about the human future, Dr. Alexander's book is a must-read."
—Whitley Strieber, author of *Communion*

"Dr. John B. Alexander has taken an unflinching, courageous look at one of the most controversial and important issues of our time. Alexander's fascinating, careful approach shifts the dialogue around UFOs to a new level."
—Larry Dossey, M.D., author of *The Power of Premonitions*

"John Alexander would make a lousy politician because he has the unpleasant habit of cutting to the chase and saying what needs to be said, regardless of the consequences. His new book will almost certainly infuriate zealots on both ends of the UFO spectrum—the diehard true believers as well as the closed-minded debunkers. And parts of it will tick off everyone in between, including me, which is undoubtedly what he intended, but whether we all agree with everything in the book or not, he is saying what needs to be said. Alexander's military mind instinctively gets to the heart of UFO cases and issues, makes quick work of charlatans and fools, and will likely inspire a new round of speculation about his presumed role as an MIB-type spook who spies on the UFO community for unknown but dastardly purposes. After knowing him for many years, I can safely predict that such speculation will delight him beyond words."   —George Knapp, Peabody Award–winning
Chief Investigative Reporter, KLAS-TV

"No one else could have written this book. You will marvel at what Colonel Alexander has done, over six decades, to be able to offer neither theories nor opinions, but facts from the real world. You will learn a lot about UFOs from this book. You will learn even more valuable information about how your government works. And you will enjoy every minute of the experience."
—Dr. Theodore Rockwell, former technical director of Admiral Rickover's
program for the nuclear navy and President Eisenhower's Atoms for
Peace, and cofounder of the engineering firm MPR Associates

"Anyone who takes more than a cursory look at the UFO phenomenon will sooner or later ask two questions: Is the phenomenon real? and if it is, What is the U.S. Government doing about it? Alexander considers both questions. His answer to the first is not surprising—yes, it is real, maybe involving extraterrestrial spacecraft. But his answer to the second is surprising—the government is not interested, and is doing nothing about it. Not everyone will agree with his answers, but most will agree that he was well qualified to look into both questions, and pursued them energetically. This book will interest anyone with an interest in UFOs—whatever they are, whatever their purpose, and whatever their origin."   —Peter Sturrock, Emeritus Professor of Applied Physics and
Emeritus Director of the Center for Space Science and Astrophysics, Stanford
University; and founding president of the Society for Scientific Exploration

"Much of the extant literature pertaining to putative UFO phenomena tends to bog down in tedious reiterations of time-worn empirical evidence and speculative theoretical propositions. In refreshing contrast, John Alexander's first-person reportage of the mish-mash of government attention (and curious lack thereof) to this potentially portentous subject quickly draws the reader into personal cognizance of an unfolding saga of all-too-familiar bureaucratic federal bumbling. The physical interpretations proposed by the author on the basis of his intimate personal involvement with this polyglot technical, political, and psychological scenario, therefore merit some careful consideration. No serious scholar of the persistent paradoxes that swirl around these bizarre phenomena should neglect these perspectives."   —R. G. Jahn, Professor and Dean, Emeritus,
School of Engineering and Applied Science, Princeton University

"Dr. John B. Alexander, Ph.D., has done it again. Perhaps even more so if one considers his previous literary and scientific efforts. Dr. Alexander is widely known for his "out of the box" thinking. This effort, *UFOs: Myths, Conspiracies, and Realities,* tops them all. Whatever you may believe about UFOs, you will be absolutely fascinated by the contents of this book." —Lieutenant General Richard G. Trefry, U.S. Army (Retired), former Military Assistant to President George H. W. Bush

"This book is a must for anyone interested in UFOs and associated phenomena. In addition to a comprehensive history, if furnishes the reader with a keen insight into the chaos that occurs when science becomes involved with politics and governmental bureaucracies. John Alexander's personal experience 'outside the box' makes this a fascinating read." —Lieutenant General Gordon Sumner, U.S. Army, ret., former ambassador at large to Central and South America

"Dr. John Alexander, Ph.D., skillfully presents the UFO question in a manner that forces one to reexamine one's own experiences and beliefs. The book is extremely well researched and touches on every aspect of what the title suggests. I am a veteran of the pioneering days of space exploration. During all my years, I never met a space professional who claimed to have ever seen anything extraterrestrial. In his book, John nails it on the unsubstantiated UFO sightings reported with not one scintilla of real evidence, yet he left on the table unanswered some reported sightings, which leaves one wondering if hidden within all the hysterical hype there might be something really out there." —Thornton D. Barnes, President, Roadrunners Internationale (an association of the CIA, Air Force, and Support at Groom Lake, Nevada, during CIA U-2 Project Aquatone and A-12 Project Oxcart/Operation Blackshield, YF-12, SR-71 Blackbirds)

"Col. Alexander's data-rich book is a must-read for anyone with even a passing curiosity about UFOs and the government's involvement in and knowledge of this controversial subject. Authored by a well-connected government insider with decades of access to senior military and civilian leaders, and who convened a highly classified, officially sanctioned UFO study group, Alexander provides a firsthand, behind-the-scenes look at the intricate government/intelligence/'military-industrial-complex' structure that emerges from his careful study. With candor and meticulous attention to detail, he shares his personal odyssey to determine the ground truth reality concerning the UFO enigma and the government's role in it. The verdict: On the one hand, he finds that our nation's best defense assets unequivocally confirm the existence of UFOs as physical objects of extraordinary performance capability, apparently intelligently controlled; on the other hand, he also deconstructs many of the field's most vaunted myths and conspiracies that have grown up around this controversial and contentious topic. Believers and skeptics alike will thus find much here to challenge a priori assumptions." —Harold E. Puthoff, Ph.D., Director, Institute for Advanced Studies at Austin

"Whether or not you are 'into' UFOs, this book could rock your world! Dr. Alexander's blockbuster book is based on about twenty-five years of searching the top levels of government and military to find 'the' group in charge of UFO information. The negative result of his search shakes the foundations of the widely held belief that there has been a government conspiracy to 'cover up' important UFO information related to UFO crashes and interactions with aliens. Anyone seriously interested in studying the UFO phenomenon should read this book."

—Bruce Maccabee, Ph.D., U.S. Navy physicist, ret.

"John's book is extremely well written and clearly lays out a compelling case that there is no central organization or committee that masterminds the 'UFO cover-up.' He's the only person I know who has the qualifications, contacts, and experience to write such a book. He puts it all together in a very readable fashion. Once you start, it's hard to put the book down."

—Colonel Charles I. Halt, ret., former Deputy Commander at Bentwaters, UK

"With his unique perspective developed over decades of cutting-edge work within the scientific, military, and intelligence communities, John Alexander offers a fresh perspective on UFO sightings and claims of government cover-up. A former U.S. Army colonel who has had direct involvement in many unusual projects, Alexander chips away at layers of false or misleading claims by overly credulous 'believers' and shoot-from-the-hip 'skeptics' as he reminds us of the importance of good sense and focuses on the evidence. He guides readers through a labyrinth of government bureaucracies and agencies as he progresses to his inescapable conclusions on the scope and depth of government cover-up and implications for society. Readers with preformed expectations may be surprised by his thesis. *UFOs: Myths, Conspiracies, and Realities* is an intriguing, wry, and unconventional book by an unconventional author. A real page-turner, it presents ideas that will either delight or infuriate almost every reader."

—Albert A. Harrison, Emeritus Professor of Psychology, University of California–Davis, author of *Starstruck: Cosmic Visions in Science, Religion, and Folklore.*

"This is one of the most important books yet published on the many mysteries associated with UFOs. It provides the most detailed and authoritative account available of an intelligent, and balanced, perspective from within the U.S. military and intelligence community. John Alexander is a skilled investigator with a careful and discerning mind capable of distinguishing fact from fiction in this most elusive and controversial of all fields."

—Jeffrey Mishlove, Ph.D., author of *The PK Man*

"Alexander provides pointed analysis on how the national government functions to address unexplained flying objects. His book is an important read on how Washington political and policy processes routinely work in examining phenomena of unknown origin."

—William H. Fite, former House Foreign Affairs Committee staff member and Executive Branch appointee

UFOs

UFOs

MYTHS, CONSPIRACIES, and REALITIES

JOHN B. ALEXANDER, PH.D.

THOMAS DUNNE BOOKS

ST. MARTIN'S PRESS ☒ NEW YORK

THOMAS DUNNE BOOKS.
An imprint of St. Martin's Press.

www.thomasdunnebooks.com
www.stmartins.com

Library of Congress Cataloging-in-Publication Data

Alexander, John B.
  UFOs : myths, conspiracies, and realities / John B. Alexander.
    p. cm.

Includes bibliographical references and index.

  ISBN 978-0-312-64834-3
  1. Unidentified flying objects—United States.   2. Military research—United States.
I. Title.
  TL789.4.A44 2011
  001.9420973—dc22

2010039157

First Edition: February 2011

10 9 8 7 6 5 4 3 2 1

This book is dedicated to all scientists who have been willing to follow the data and acknowledge that phenomena do exist—even at the risk of their careers.

On July 20, 2010, Dr. Dean Judd died unexpectedly, and thus I can now publicly acknowledge his involvement. For many years he was one of my closest confidants and his support is evident in many places in this book. Dean served the nation as the Chief Scientist of SDI (Star Wars), the National Intelligence Officer for Science and Technology, a member of the U.S. Air Force Science Advisory Board, and Senior Fellow at Los Alamos National Lab. He was a true skeptic. We served together on several advisory boards, and often discussed matters on national security at great length. On UFOs he asked tough questions while providing key insights into the U.S. space architecture. He was one of many people who noted there was absolutely no connection between SDI and preparing to fight ET.

Finally, I thank my wife, Victoria, who sometimes bridges realities with courage that would fail most people.

CONTENTS

Acknowledgments   xiii

Foreword by Jacques F. Vallee   xv

Commentary by Tom Clancy   xix

Introduction by Burt Rutan   xxi

Prologue   1

1. Advanced Theoretical Physics Project Disclosed   6

2. The Corso Conundrum   40

3. The Condon Report Revisited   51

4. The Congressional Hearings That Never Happened   65

5. The Government and UFOs   75

6. The Presidential Paradox   101

7. Majestic 12   122

8. The Apollo Lunar Program   134

9. Reverse-engineering UFOs   144

10. Real Cases and Hard Data   155

11. The UK Connection   183

12. Sworn to Secrecy   196

13. What NATO Knows   212

14. Considerations, Speculation, and Puzzles Addressed   222

15. The Twilight Zone   255

Epilogue: Quo Vadis?   269

Appendix A   275

Appendix B   285

Appendix C   287

Appendix D   294

Index   297

ACKNOWLEDGMENTS

Over the decades many people have supported my efforts in researching UFOs and other phenomena. For good reason, some of them choose to help from the sidelines and remain anonymous. That is true for several members who participated in the Advanced Theoretical Physics group. Very importantly, my acknowledgment does not mean that we necessarily agree on all aspects of the topic—and in some cases we profoundly disagree.

Among those who deserve thanks for participation and support are Hal Puthoff, Kit Green, Jacques Vallee, Edward Teller, Tom Clancy, Bob Jahn, Brenda Dunne, Peter Sturrock, Ted Rockwell, Ben Rich, Scott Jones, Jack Houck, Howell McConnell, Marty Piltch, John Mack, Dave Pritchard, Bill Coleman, Phil Corso, Edgar Mitchell, Ron Blackburn, Bob and Ryan Wood, Eric Davis, Colm Kelleher, George Onet, John Petersen, Jim Channon, Max Thurman, Dick Trefry, Gordon Sumner, Frank Burns, Bob Klaus, Tyler Nelson, George Knapp, Al Harrison, Burt Rutan, Jack Schmitt, Bert Stubblebine, Bruce Maccabee, John Schuessler, Chuck Halt, Bob Salas, Lynne Kitei, Nick Pope, Michael Hesemann, Illobrand von Ludwiger, Anne and Whitley Strieber, Roger Leir, Steve Bassett, Gordon Novel, Helen and Bob Kupperman, Bob Friend, Paul Tyler, Kreskin, Joe McMoneagle, Ingo Swann, Paul Smith, Bill McGarity, Oke Shannon, Stan Friedman, Larry Dossey, Mark Rodeghier, James Gilliland, Gabe Valdez, and others I may have inadvertently left out. My appreciation also goes to my family members who provided details of their sightings.

Special thanks go to Bob Bigelow for the creation of the National Institute for Discovery Science and allowing me the opportunity to participate for several years.

Finally, thanks also go to my editor, Pete Wolverton, and his staff at Thomas Dunne Books for their support on this project.

The 1980s were a time of intense technological advance and saw much speculation about frontier scientific topics, ranging from parapsychology and virtual reality devices to the search for extraterrestrial intelligence and the nature
of unidentified flying objects. The U.S. Government was involved, or ru-
mored to be involved, in many of these explorations; and in the context of
these projects one name kept coming up, that of Dr. John Alexander.

A Green Beret veteran with distinguished service in Southeast Asia, Dr.
Alexander has served in high-level management positions since he retired
from the Army with the grade of colonel. His assignments have brought him
into contact with some of the highest scientific and government circles in the
United States and Europe.

In October 1989 Colonel Alexander came to see me in my office on Sand
Hill Road in Silicon Valley. It was a meeting I am not likely to ever forget. We
were going to discuss our favorite subjects: remote viewing research and the
latest experiments John was conducting in biological correlations to distant
influence. We also planned to compare notes on unidentified flying objects, a
topic of frequent discussion among high-technology entrepreneurs with
whom I worked and networked in California: understandably so, since the apparent behavior of these objects raised fundamental issues about physics, advanced materials, and physiological reactions among human witnesses. John
had organized a serious, multiagency task force, intending to analyze the
massive data accumulated by the U.S. military regarding the phenomena, but
it had run into bureaucratic obstacles. In spite of active interest in some parts
of the government, the stigma of the paranormal proved too much for the
Pentagon culture to overcome, even in the rarefied atmosphere of a classified
study; which reopened the obvious question: Was there an even more secret
study, somewhere else?

I can testify that many senior people in foreign countries (particularly in
Europe) felt certain there was such a study within the United States, although

it was never clear what the source was for this strongly held belief, which continues today. A recent French documentary on UFO cases around the world came out with the silly title *The American Secret*, even though it did not provide any indication that there was such a secret. I met a similar belief among American agencies with which I had the opportunity to interact in my professional role as one of the principal investigators in the early days of the Arpanet but, surprisingly, they always assumed the secret project must be located inside one of the other agencies, to which they had no access! In his subsequent study, Dr. Alexander made the same observation, and the question remains very much open today.

Whether or not there is a secret study somewhere, the fact is that first-level sightings of unidentified aerial phenomena have been routinely denied, ridiculed, or withheld by military authorities in this and other countries. In spite of all the obstacles thrown in the path of serious investigation there were scientists, entrepreneurs, and military officers willing to risk their careers to help make sense of the bewildering data that came from pilots, control tower operators, foreign observers, and even from the most deeply classified assets in the air and in space, and Dr. Alexander was one of them.

Unfortunately, John and I never got to the most interesting topics we had planned to discuss on the day of his Menlo Park visit. Our conversation was brutally interrupted, not by bureaucratic interference or by angry skeptics but by a genuine Act of God: My office started shaking. My windows, which framed a beautiful view of the coastal range across the San Andreas Fault less than a mile way, actually buckled. So did the floor. John got up from his chair and smartly braced himself in the doorway, while I looked for shelter under my desk, as every responsible Californian has trained himself to do.

The Loma Prieta quake was a long one and it seemed that the building itself might start to come apart, but things eventually returned to normal in the room. Naturally, John and I worried about our families and friends and had other immediate things on our minds than the reality of UFOs, so we left and drove away with very little protocol. It will be recalled that this particular quake damaged and closed down the Bay Bridge, killed nearly a hundred people under collapsed highways, interrupted power distribution for days in San Francisco, caused billions of dollars of damage, and disrupted life for a year in Northern California.

Any extensive conversation on "our subject" was on hold for many months, until John and I had the opportunity to discuss it again as space entrepreneur Robert Bigelow organized the National Institute for Discovery Science in Las Vegas, Nevada. I learned that in the course of his work within the military, and in his subsequent efforts to network with interested civilian experts, Colonel Alexander had obtained important new information on the phenomenon and the social, cultural, and political forces that surround it. At the time he was too busy to compile the data in a form suitable for wide publication, so only fragments were picked up by the media and immediately transformed into colorful but misleading rumors.

We are fortunate that Dr. Alexander has now assembled his observations and conclusions in this important book. He does not claim to answer the question of the origin and nature of the UFO phenomenon but he cuts through the ambiguity that surrounds it, highlighting the real issues among a blizzard of theories and false claims. If a serious study of the subject is ever undertaken, it will find in this volume much constructive information (including previously classified facts) and a wealth of remarkable insights. Some of the statements from prior research should be debated in a climate of open, rational exchange: Is there really no hidden study, within the government or elsewhere? Is it true that UFOs do not represent a threat to our national security? And perhaps most important, how long will scientific opinion continue to deny the reality of a phenomenon at once so complex and so rich in opportunities for good physics research?

John Alexander's book touches on all these topics, citing a clear, timely, and thoroughly documented array of statements and observations. Other researchers may disagree (as I do) with some of his specific conclusions, but he does present evidence for a massive failure in our collective ability to deal with many unexplained phenomena that surround us. In that unique sense, his book is a major contribution to the cultural debate about the future of the human race as it tries to come to grips with the complexities of life and consciousness in the universe.

Jacques F. Vallee
San Francisco, August 10, 2010

COMMENTARY

Like most Americans, I've been reading about UFOs since I saw my first flying saucer movie back in 1954 or so. First with the innocence and willingness
to believe of a kid, but as I grew up, so did my skeptical nature, and along the
way I just wanted to find a source of information which clearly had some acts
and reason behind it. It's finally available. John Alexander is a serious professional soldier with a Ph.D. behind his name, and with work behind him at a
classified national laboratory. He also has an active, questing mind and is always out looking for information.

This book is, I think, maybe the first-ever look into a question that millions
of people have, and more to the point, it's a look that asks questions and looks
for hard answers. John, whom I met back in the 1980s, is a guy who knows what
to look for, knows whom to ask, and knows the language of the answers to his
questions. What he says you can take to the bank. He also knows what he cannot say, and in some ways the book is as frustrating as many of the "I know"
UFO books. The final word is not yet in, even here. You won't find any of the
"I know but I can't say" stuff. Sources are identified, and can give the reader a
place to start his own inquiry.

UFO lore is largely a place where history and science meet theology. People believe because they want to believe. But they want to believe in what?
Arthur C. Clarke once said that any sufficiently advanced technology cannot
be distinguished from magic. We know that's true. I can remember what
computers were back in the 1950s—big adding machines. In 1984 I saw the
first Mac unveiled in D.C. and decided on the spot that I had to have one. I
could not write books without them, and they're as common as rainwater, and
almost as cheap. But fifty years ago, they had not been imagined. Forty years
ago, it took an IBM System 360 that filled a room. Twenty years ago, the first
portable came out, and now the important part of a portable computer is the
keyboard you use to dial the phone. Now, had one invented an Apple Macin-
tosh back in 1910, the hard part would have been not to be burned as a witch.

We can assume that UFOs, if they exist, have computer power such as we have not yet dreamed about, and what else can they do? Dimensional corridors that bypass space-time? If so, then they can hop from place to place faster than the speed of light. Or even time travel? What we consider reality would be immensely limiting to the pilots of such craft.

We cannot know, and even if we did know, could we understand?

Probably, yes.

We can learn damned near anything, but learning takes time. Consider a cure for cancer. Some very smart people are working on that. Some of them are friends of mine. I cannot grasp what they are doing, but when they are successful, it will in retrospect appear simple. That is the way of progress. It is the nature of science to make the incomprehensible transparently obvious. Just that you cannot know what you have not yet discovered.

And so it is with the UFO. We do not know how they, if they are real, operate. All we can do is to gather information, draw conclusions from it, then advance a step at a time in the hope of catching up. As with the growth of a child, life is about learning and categorizing reality. It requires time and patience and thought. John's book is a step toward the learning. It outlines what evidence we have. Perhaps in retrospect we will see how easy it ought to have been, but you need to walk forward before you turn to look back. When Magellan sailed west from Portugal, he didn't know what he would find, but for him that was the fun part. John's new book is part of Magellan's log. Maybe he doesn't answer all the questions, but he sees more than anyone else I know of has. So, he has taken a step. It's our job to learn from him in order to take a few more.

Tom Clancy
Baltimore, 2010

First, I want to thank my friend John Alexander for inviting me to submit my thoughts for publication in his excellent book. In general, I never agree to provide comments on any books I have not read, and I have always found it difficult to read any book on UFOs. They all seemed to be written only by those strongly advocating their chosen side of the debate of the big question "Do ETs/UFOs exist?" These books would be compelling if there were convincing physical evidence to inspect (hardware, bodies, wreckage, etc.). My impression usually is that there is no compelling evidence and the author's credible data could be summarized into a few pages. However, they then seem content to drag the reader through hordes of speculation and rhetoric for no useful purpose. Thus, I have to admit I had not read an entire book on the subject before John provided me with his early draft of *UFOs: Myths, Conspiracies, and Realities*. John's book is very different; its detailed coverage of both sides of the debate is written without bias or agenda.

A Skeptic

John has rightly described me as a skeptic. This is a refreshing truth, especially after *Wired* magazine had printed that I "think Extraterrestrial Aliens built the Gaza Pyramids"! During a long interview for the magazine on a different subject I included thoughts from my study of the Egyptian monuments. I mentioned my conclusion that several of the structures had clear evidence of manufacturing technologies well beyond what mainstream historians assume existed at the time. This is a fascinating subject indeed, but it has nothing to do with UFOs. For me, it is far easier to believe that the ancients had developed methods to cast granite and that these methods were later forgotten, than to believe that starships had visited Earth to help humans stack stones.

Since college graduation in 1965 I have always been involved in research, development, and flight-testing of aircraft. The first seven years were spent as a U.S. Air Force flight-test project engineer at Edwards Air Force Base, and the next ten years as a homebuilt-aircraft entrepreneur in the Royal Air Force. The last twenty-eight years I built a small aerospace company, Scaled Composites, of which I'm the founder and CEO. These efforts brought me in contact with much of the most advanced research in aviation.

My Egyptian pyramid study was just a hobby; I have always enjoyed researching some kind of mystery to provide a welcome distraction from my day job. I do not claim to have solved any of the mysteries; I merely dig away to get to the point where my own gut says, "Now I know." This is not a proof; it's merely satisfaction in my own mind that further study will be boring, thus forcing me to find another hobby.

My hobbies have included:

- Energy efficiency (I built solar water heaters in the 1970s; I built my custom home in the 1980s, which *Popular Science* called the "ultimate energy-efficient house"; I drove an EV-1 electric car for seven years as my primary vehicle' and I am currently working on a twenty-acre PV solar farm).
- Investigating the JFK assassination; my fascinating search to solve the murder mystery started in 1990 and ended in the late 1990s, when my gut told me I probably had found the answer.
- My pyramid-manufacturing quest; a hobby starting in early 1998, when I spent several weeks looking at structural details (with an engineering/manufacturing eye, rather than reading hieroglyphs) from Cairo to Abu Simbel.
- My current hobby (since 2006) is a survey of climate data and an assessment of the analysis and presentation methods of those promoting the theory of catastrophic global warming caused by human emissions. Like my earlier hobbies, I will likely soon move on to another one when I am satisfied that I know the truth, and then get bored. When I am finished with a hobby I tend to never look back.

My UFO Sighting

Even though I have seen a flying object that I could not identify, I have never made it a hobby to study UFOs. However, the field keeps coming to me. This is my first-ever publication of my 1972 sighting. It occurred while I was driving northbound on I-135 in Kansas by myself in the dark, predawn hours of May 31, 1972. I was on my way to direct the first flight test of the BD-5 when off to the right I noticed a brilliant, hovering, cylinder-shaped object with a length about four or five times its diameter. Assuming it was not moving, it was easy to determine its size and its distance from me by knowing my car's speed and the angular rate at which my line of sight rotated from the center windshield to the right side window. I slowed to stop on the side of the road and when I had stopped the object quickly accelerated to the south and disappeared within two seconds. I estimate that it was about 300 to 400 meters away and its size was about 30 to 50 meters long. It had been hovering at about 10 to 20 meters above the ground.

It did not appear to have a solid surface; it was more like a fuzzy fluff of glowing light with no distinct shape or surface details. During the approximately fifteen seconds that it was visible, it changed colors at least twice; green, orange, yellow, as I recall. I took note of the odometer so I could find the site again, and when I returned in daylight I found high-voltage electric transmission lines crossing a small lake near the spot where it had been hovering. I therefore formed an opinion that the "object" might have been some electrical ion phenomena. However, I was surprised to read a newspaper account that had a description of a very similar object sighted about two weeks later in Louisiana.

My use of a *Men in Black* neuralizer on the audience at one of my Oshkosh/ Air-Venture talks in 1997 led some people to think I am an ET/UFO believer. Engineering cohort Dan Kreigh, using a toothbrush case and the components of a flash camera, built the device. Yes, it was only a joke; humor is needed when you work in the high desert. Dan was also the one who painted an alien face on the back window of my Boomerang twin aircraft.

Another reason that people might think I am a believer: Some of my personal friends are among the people that are described in John's book:

- Ben Rich. The Lockheed Skunk Works of the forties, fifties, sixties, and seventies have always fascinated me with their methods of aircraft research, and I have strived to apply some of their lessons-learned to my companies. The Skunk Works was founded on June 17, 1943, the day I was born. I did get a chance to briefly meet Kelly Johnson in the 1970s. Ben Rich and I became friends after I met him at a government technical conference. We often discussed advanced aircraft programs but he never even hinted about the ET subject.
- Lieutenant Colonel Phil Corso. I did attempt to read his book *The Day After Roswell*, but not because of his ET claims. My interest was the fact that in 1964 he was assigned to Warren

Commission member Senator Richard Russell, Jr., as an investigator into the assassination of JFK. I managed to locate him via his son who was employed in the same business as my RAF homebuilt aircraft company. In early 1998, during a family vacation at Disney World, I took a day off and drove to the Corso home to interview Phil. I managed to find him alone and talked to him for several hours. Phil was a very different person to talk to alone than when his family was with him. He seemed sincere and open while alone, but guarded and controlled when his son was present. He seemed surprised and happy that I was interested only in his Warren Commission work. He provided some remarkable teasers about the JFK case but said he would "only elaborate during our next meeting, not today." I pressed him hard but had to leave and of course had planned to meet him again. However, he died soon after my first and only visit. I later asked his son for any documents, any draft or notes he might have written on his other book, *The Day After Dallas*, but he refused the request. Overall, I believe Phil was a real gentleman. He was fascinating and fun to talk with. But I did not get a strong feeling that his ET or reverse-engineering information was credible.

• Hal Puthoff, John Alexander, and Bob Bigelow. I consider these folks good friends and I have enjoyed my interface with them for many years. I am very interested in their work, their passion, and their capability. However, I would not have met these folks when I did were it not for an introduction by someone who is an ET/UFO true believer. He approached me with verbal technical proposals for wild, far-out propulsion-development products, but I never officially engaged him on those subjects. My incentive to return his calls was solely because my then-current hobby was studying the JFK case. The guy's reputation included a long history of involvement in the JFK presidency before and after the assassination in 1963. I did not learn anything new or interesting from him; however, I do owe him a big thank-you for introducing me to Hal and then to John. I also want to thank John for introducing me to Bob. I am very interested in Hal's work on future energy systems, John's work on nonlethal weapon systems, and Bob's impressive, groundbreaking work on developing nongovernment space stations. However, I have never been a NIDS participant and my friendship with these three folks has nothing to do with the ET lore.

- Wernher von Braun. I met von Braun in 1965, in San Francisco, where we both were receiving AIAA awards. I chatted with him at the cocktail hour, but only about his Apollo work. My closest golfing friend started his NASA career by flying von Braun around the country in the 1960s, and among other friends was close to von Braun at the Marshall Center in Huntsville, Alabama, until he departed for Washington D.C., in 1970. None of these people ever mentioned to me a hint that von Braun was aware of, or was working on, any ET issues.

Bottom Line

A reason that I remain an ET skeptic is that, for forty-five years I have been in a position to handle sensitive technical information and have not heard of anything related to ET hardware or reverse-engineering projects. I have served two separate five-year terms on the U.S. Air Force Scientific Advisory Board, including a group assigned to study and evaluate research for advanced propulsion systems (at Wright-Patterson Air Force Base, the Air Force Research Laboratory Propulsion Directorate, and at other locations), but I have never heard anyone mention the kind of stuff that excites the ET/UFO crowd. I hear only the wild claims from people that promote the lore, but never from any government official or other credible source.

Finally, claims that ET spaceships have crashed numerous times worldwide since the 1940s, without any portions of wreckage ever being available for nongovernment scrutiny, and without any released or leaked government analysis or photos of wreckage, just does not pass a sanity check. In that vein I concur with the analysis and common sense John presents in this book. Until convincing physical evidence is found, I will remain a skeptic of such assertions.

Lookin up . . . way up,
Burt Rutan
Founder, CTO/Designer Emeritus of Scaled Composites;
winner of the Ansari X Prize and the Charles A. Lindbergh Award;
Time magazine "100 Most Influential People in the World"

UFOs

Imagination is more important than knowledge.
—ALBERT EINSTEIN

UFOs are real! With no prevarication or qualification of terms, there are physical objects of unknown origin that do transit our universe. The evidence that supports those statements is simply overwhelming. That evidence includes both hard data collected via multispectral sensors and from high-quality eye-witnesses that are neither misreporting facts nor delusional. Determining what these objects are, let alone the question of origin, is another matter. There are no simple solutions that fit all of the facts. For about six decades I have been actively exploring the topic and I'm finally taking the time to write about it.

This book differs from all others written about UFOs because it details a personal exploration of the topic in ways that no other researcher has been able to accomplish. What is important and unique are the details of direct interactions with senior officials from many agencies—and finding that no one was responsible for this topic anywhere in the U.S. Government. That fact is very significant and runs counter to conventional wisdom. This book addresses those problems so that the reader can understand the government's role in UFOs as it really is; not the way many people believe it should be.

Requirements for Researching UFOs

In presentations and discussions on the topic of unidentified flying objects (UFOs), I always begin by stating my position. This is important as there is considerable misinformation about me and my role in the investigation of this fascinating subject. There are three things that anyone who wants to become involved in the study of UFOs will require. While they will be more fully explained later, these attributes are:

> — First, a sense of humor; as you will be attacked sooner or later.
> — Second, an understanding of conspiracy theory; you are now part of the plot.

— And finally, a day job—or be independently wealthy;
nobody is making big money from the topic.

My position is simple: by definition, UFOs must exist. At the most elemental
level of understanding, this means that any object that flies, and is not identified, is a UFO. Of course many things begin as UFOs, and are later identified
as some prosaic alternative, and thus move out of the UFO status. But that is
not what you're here for. What we are really discussing are mysterious craft
that appear in physical reality and that have been observed and reported for mil-
lennia. Most important, the evidence supporting the reality of UFOs, as they
are commonly understood by the general public, is proven beyond any reasonable doubt. There is a core set of cases—that after you eliminate all of the alternative explanations—just defy conventional wisdom. Any so-called skeptic
that states, "There is no hard evidence to support the existence of UFOs," is
simply wrong! What he or she really means is that they haven't taken the time to
seriously review the data, and most likely you can infer that they aren't about to
start now.

Over the past decades I have had the opportunity to research the topic both
officially and as an avocation. My interest in UFOs began with the first public
pronouncements of Ken Arnold's experience and various other sightings in
1947. When I was ten years old, I remember a broadcast of the topic on our
school's internal radio system. I attended the Training School, which was located on the campus of what was to become the University of Wisconsin–La
Crosse. As a side note, the name of the school was changed as some people
thought we went to a reformatory, rather than a very progressive school that
provided human guinea pigs for young student teachers.

Throughout the course of this book what I have learned will be covered in
detail. I promise to tell you only things that are believed to be true. There will
be no intentional misleading of the reader. It must also be accepted that people
can, in good conscience, look at the same facts and arrive at different conclusions. In fact, I have some good friends whose opinions are highly respected
that fall into that sector. Later some of those alternative opinions will be addressed specifically as well as the reasons that we see the conclusions quite
differently.

Frequently I'm asked whether or not I've seen a UFO. The answer is yes,
but it was not very dramatic. When visiting James Gilliland at his remote site
in Washington State, well known for sightings, I did witness a minor unique
event. This nighttime observation included what appeared to be a satellite moving across the sky from west to east. I was watching this high-altitude object

through very good, third-generation night-vision goggles when it suddenly stopped moving. Satellites don't do that. Compared to what others have reported, that is hardly earth-shattering, but it's the best I have been able to come up with. More important, my studies indicate that while personal observations are interesting, it is the multisensor data that have been reported by many different systems that confirm the reality of physical craft beyond any reasonable doubt.

Who Do You Believe?

There have been some observations of UFOs in my immediate family. My brother Don told me about two sightings, one of which was confirmed by my other brother George. One he described as follows: "It was in the summer of 1958 in west Miami, Florida, around 2000 to 2030 hours. I was with several friends and we spotted several cylindrical shapes playing in the sky. They moved slowly and methodically in formation, coming in from the northwest sky and traveling in a westerly direction. There were three or four of these objects floating across the sky. We watched them for a short time from in front of [a friend's] house when we heard the sound of jet engines from aircraft. We then saw three jet fighters starting to track after the objects, which in turn put on a blazing burst of speed and fled the area—in leaving they did not fly in a straight line, but made several abrupt angled turns and within seconds were out of sight."

Don had an earlier sighting that took place in La Crosse, Wisconsin, before the days of satellites circling the globe. He stated, "We were looking up as kids do, when we saw two lights in the sky that were moving from east to west and were very bright in appearance. When they did a turn to the south, it was not a normal graduating turn like an airplane, but a sharp right-angle turn. We watched it for a while. It was coming toward our direction in the sky when it made a sharp turn, put on a burst of speed, and was gone in seconds."

Similarly my son Mark, now a lieutenant with the Palm Beach County Sheriff's Office in Florida, has told me about two sightings he had. One was while fishing one evening in his small boat in the Atlantic Ocean off the Florida coast. There he saw a bright object that moved across the entire sky, but did not behave as a meteor would. The other incident involved a UFO in close proximity to the ground, which resulted in a massive power outage in the western residential area of the county that encroaches on the Everglades. When the electricity went off, Mark left his house. At that point he saw the UFO that had been hovering shoot rapidly straight upward and disappear into the night sky.

As UFO sightings go these are not terribly unique or remarkable. However they do have two important qualities. First, like most sightings, the witnesses never reported the event to any agency. Rather, they remain tucked away as family lore. But that is the second point: they do come from a trusted source, one who has no reason to lie. There are many stories about UFOs that are hard to believe and where the source's credibility is unknown. However, if you cannot believe your own family, then who can you trust? That is a theme often repeated as people relay relatively unknown accounts of personal observations.

There are those in the conspiracy theory crowd that believe that I'm part of MJ-12, or some other mystical group that is part of a sophisticated cover-up scheme. Depending on which conspiracy one believes in, this cover-up is run by the U.S. Government; possibly some rogue element of a black organization, or an international cabal controlled by the Bilderbergers, Trilateralists, or the Council on Foreign Relations. Many conspiracy theorists (CT) just seem to believe that there exists some great unidentified *THEY*. This is not true. I'm not part of any covert group and I will have much more to say about conspiracy theories later. Of course, in the convoluted logic of that same CT crowd, just denying association is proof that *THEY* exist.

I will discuss in some detail an ad hoc, interagency group that I formed explicitly to explore the subject of UFOs. Part of that journey found me briefing quite a few very senior people and their responses were quite revealing. It can be safely said that what was found was not what was expected at the beginning. While many researchers, and even the general public, indicate they believe the U.S. Government is withholding critical information about UFOs; that is not a position in which I concur. The vast majority of the independent UFO researchers who believe in cover-up and conspiracy theories simply do not understand how the government works. This is the greatest shortfall that I have found with the conclusions they have drawn. Hopefully some light will be shed on how macro-systems function and why understanding that social psychology applies to the UFO field just as much as in any other endeavor. Despite what you may think, UFOs are not an exception to the rules and norms of social sciences.

Whenever I give a presentation on phenomena I list a set of my obligations. They apply in this book as well. Those rules are:

— I will tell the truth as I know it
— There will be no intentional disinformation
— There are things I will not tell you (Because they are
    formally classified)

The information provided is as straightforward as I know it. I still hold security clearances and am bound by the applicable laws. However, I have never encountered a situation regarding UFOs that came close to divulging classified information. Security and classification issues will be discussed in a separate chapter. What most people think about those issues is wrong. Throughout the book people who have agreed to be identified, or have already exposed themselves in other fora, will be named. Anyone who has died will be named. There are people who were engaged in various projects whose identity I agreed to protect at that time. I will continue to honor that agreement.

It is acknowledged that there are those critics who think I just missed the mark and was lied to, or intentionally misled, by those senior officials with whom I interacted. While that is possible, I believe it to be highly improbable because of the number of people involved and my relationships with them.

Whatever UFOs Are, the Answers of Origin Are Not Simple

For the record, I'm convinced that there is no single concise explanation that fits all of the data. Whatever we are looking at, the epistemology is terribly complex, some of the events are very real yet may have a temporal component, and they present elusive attributes that perplex even the best minds. All that is asked is that you review the material and apply common sense to your evaluation. If there is one ingredient most lacking in UFO studies it is that—common sense.

First we will explore how this internal ad hoc UFO investigation came about. There were many moving parts and understanding how they all came together is a bit complex. However, to comprehend the significance of our conclusions it is important to have some of the background elucidated. Note that synchronicity places a role that should not be ignored. Key people kept emerging, often in unique or unanticipated ways. The project covered next has never before been revealed in writing by a participant.

1

ADVANCED THEORETICAL PHYSICS PROJECT DISCLOSED

Do or do not . . . there is no try.
—MASTER YODA, *THE EMPIRE STRIKES BACK*

"There is a black UFO program someplace. Let's see if we can find it. If the rumors are right, THEY have a tiger by the tail and want help transitioning the information to the public. Somebody has got to be minding the store—who is it?"

It was with that premise that the *Advanced Theoretical Physics Project* (ATP) was born. I'd like to say it was in the bowels of the Pentagon; it has such a nice ring to it, but that would not be true. Rather, it was located in my unbelievably spacious office in Tysons Corner, Virginia, with views of the Shenandoah Mountains to the west.

With help from a few other researchers we initiated the ATP project specifically to examine issues regarding UFOs, and what role the Department of Defense (DoD) might be playing. Like most of the general public, we assumed somebody in DoD had responsibility for studying UFOs. This was a small internal group drawn from the government and aerospace industry, all of whom were interested in the topic. Admission was by invitation only, and those selected to participate had to have the right credentials. This chapter outlines what few people outside the membership of that group have ever heard about. In fact, very few of them know the full story. What we expected was not what we found.

Years later, sitting in front of Lieutenant General Jim Abrahamson, then the director of the highly acclaimed Strategic Defense Initiative (SDI), known more popularly as Star Wars, I began the briefing to him and explained that we were there to discuss UFOs. It took only a few minutes before he stopped me and asked, "Who are you guys really?" That spoke volumes about the subject and where UFOs were on the SDI priority list. This is the story of how we got there, and beyond.

To appreciate the importance of the research presented, background material is provided to allow better understanding of the depth of penetration that was accomplished. Of significance are the personal relationships that intertwine over time; always assisting when needed. Note that the interactions with most agencies were not simple onetime interviews, but ones that gave continuing support at various levels. Also of importance was the manner in which traditional military research and development functions were juxtaposed with UFO interests. Understanding how that unique fusion occurred is critical to knowing how this effort was accomplished. It was that aspect that separates this research from all others in the field.

Setting the Stage

The creation and execution of the Advanced Theoretical Physics group did not just happen. Rather, it took considerable coordination and effort. Often the direction I was moving in my career was not clear to me at the onset. In fact, at my retirement from the Army, I told the assembled group that it was as if some external force seemed to guide my actions. When I resisted various movements and failed to get my way; it was only later that the necessity for certain events to have occurred become apparent. Maybe it was synchronicity, but the eventual functions and outcomes that led to the observations in this book were far too complex to have been planned or orchestrated by me. While the emphasis here is on UFO studies, there were interactions in many other realms of phenomena that remain totally intertwined with destiny.

To understand the significance of this research effort, some fairly extensive background knowledge is required. In that way the reader will be able to discern how all of the pieces fit together. The ATP project was actually the result of substantial groundwork in establishing various credentials in the military that allowed me to function within acceptable limits while simultaneously pushing the boundaries of traditional credibility. It also required finding the right network of people who would be willing to participate in projects that could be perilous to one's career, and establishing the set of senior officials who could fly cover for such an operation. In general, the military is fairly risk-aversive. To get promoted in the military the conventional wisdom was to assume assignments where the potential for failure was minimal. Venturing into studies of psi phenomena and UFOs was not seen as career enhancing. As a mustang (former enlisted noncommissioned officer who went to Officer Candidate School), and who had chosen Special Forces missions over straight infantry assignments, I knew my time was limited and further promotion beyond lieutenant colonel was out of the question—but that was wrong.

Enter the Voodoo Warriors

In the summer of 1980, recently promoted and fresh out of Command and General Staff College (CGSC) at Fort Leavenworth, Kansas, I was assigned to the Headquarters of the Department of the Army Inspector General's office (DAIG). There we conducted inspections on systemic issues as selected by the Inspector General of the Army, Lieutenant General Richard Trefry.

While attending CGSC I had written and submitted a manuscript to *Military Review,* the staid journal of the college. The piece was called *"The New Mental Battlefield."* This article described the potential use of remote viewing, psychokinesis, and similar psychic skills in military operations. In fact, this was the first time that these subjects had ever been broached in an official U.S. Army publication. The article was published in December 1980 and, strangely, became the cover story for that issue. In one of many synchronistic incidents, the editor of the magazine had previously had a near-death experience in which he found himself out of his body. He was therefore fascinated with the article, and the cover came complete with pictures of Kirlian photography. While they were not really germane to the story, this was quite a departure for a traditional professional military journal. The immediate impact was relatively muted within the Army and I continued with regular IG work. However, Ron McCrae, one of the staff writers for the noted Washington columnist Jack Anderson, spotted the article. McCrae drafted an editorial for Anderson which was then published in *The Washington Post* under the title *"The Voodoo Warriors of the Pentagon."*

Obviously Jack Anderson's article, which was picked up by the wire services, was not very supportive of the notion of military applications of psychic capabilities. Anderson was known for finding topics that would incense the public, and often dealt with government waste of taxpayer money. Certainly, the writers assumed, the concept of government funding psychic warfare would get a response from the public. Anderson and McCrae were right, but not as they had anticipated. Rather than being angry about sinking money into a possible psychic warfare program, the public wrote in to tell Anderson how much they appreciated knowing that this research was going on. Given the 1970 publication of Sheila Ostrander and Lynn Schroeder's popular book, *Psychic Discoveries Behind the Iron Curtain*, many civilians believed that the Soviets were seriously pursuing these topics and that the United States had fallen behind.

It was from Anderson's article that I experienced blowback as well. One morning, while sitting at my desk in C-Ring on the first floor of the Pentagon, a red-faced lieutenant colonel accosted me. He was assigned to the Office of the Army's Assistant Chief of Staff for Intelligence, Major General Ed Thompson, and they were furious about the Anderson exposé. He first asked if I had written the article, and then he wanted to know if I was aware of the clearance procedures for publications of any nature. Fortuitously, I happened to have the file with me that contained all of the administrative paperwork associated with the *Military Review* article. Once shown that his office had signed off approving the publication of the manuscript, he left in a huff. Of course what had upset them so much was that they had responsibility for the secret embryonic remote viewing program, then-called *Grill Flame*. After many years in the black world, this program was declassified in 1995, and became known to the public by its final designation, *Star Gate*.

Task Force Delta—Defining the Difference that Makes a Difference

Although officially working for the Inspector General of the Army I was already participating in a very unique organization called Task Force Delta. This is not to be confused with the more famous Delta Force, à la Colonel Charlie Beckwith, with a counterterrorist mission. (As a side note, "Charging" Charlie Beckwith and I were students in the same Ranger class [2-58] in the late summer and fall of 1958. He was a decidedly boisterous lieutenant, and I was a very junior enlisted soldier who was promoted to sergeant by graduating from that grueling course.)

In fact, this new Task Force Delta was an Army version of a think tank and was facilitated by some very enlightened senior leadership. That top echelon comprised three- and four-star generals who knew the Army needed dramatic revamping given the vicissitudes that followed Vietnam. There was a need, they thought, to have people available to address tough, multifarious problems in an unconstrained environment. Blue sky thinking was not only allowed, it was encouraged.

It was there that Lieutenant Colonel Jim Channon brought out his *First Earth Battalion*, a notional unit with concepts easily three decades ahead of their time. Channon's project would become better known to the public when Jeff Bridges played a distinctive character based exclusively on him in the 2009 movie, *The Men Who Stare at Goats*. A brilliant imagineer and craftsman, Channon was the artist of choice for many senior flag officers in the Army Headquarters. He had a unique ability to listen to generic ideas, and then graphically portray them with a stunning mastery to convey complex subjects so that they had clarity and were understood by even the most adamantine staff officers. Channon was one of a kind: intelligent, intuitive, caring, and extremely gifted. By traditional Army standards, while he could bridge both worlds, many of his fellow officers would think him a bit too unorthodox for the conventional force. That never stopped Jim and today many of his basic designs have been inculcated into standard Army graphics. As a result of cross-pollination, his drawings included translating some of my advanced concepts into a visual format for briefings at the highest levels in the Army. We remain friends to this day.

Task Force Delta was then led by Colonel Frank Burns, an innovative organizational effectiveness officer with extensive training in neurolinguistic programming (NLP). There were only five people permanently assigned to the cell, and three were strictly administrative. Burns's deputy was a civilian, Tom Kelly, who in later years would become the Deputy Assistant Secretary of the Army during the George W. Bush administration. Thanks to Burns and Kelly, members of Task Force Delta were employing information technology over phone lines long before the Internet was up and running. Current computer users, with downloads measured in megabytes, probably cannot even imagine having documents transmitted at 600 baud per second. At that slow rate we could watch each letter as it was printed out—and graphics were unheard of.

Other than the five-person core group, participation was purely voluntary. Some of those who participated went on to lead the Army, and at least one, Colin Powell, became a household name for his achievements. While predominantly Army personnel played, we also had a fair-sized group of civilians who were sufficiently intrigued that they paid their own expenses to attend meetings that were held on a quarterly basis. Task Force Delta was an organization that thought so far outside of the box that its members didn't know whether or not a box existed.

For me, the important aspect of Task Force Delta was the people with whom I came in contact both inside and outside the Army. The free exchange of ideas at times included informal discussions of UFOs—but I learned who was open to such discussions, and who was not.

One of those who would consider unusual topics was the Deputy Undersecretary of Defense, Richard G. Stilwell, a retired Army four-star general, who had been appointed in a civilian position as the number two person in the Department of Defense. In reality, it was General Stilwell's wife who suggested that he contact me and arrange for a meeting on phenomenology. As a lieutenant colonel assigned in the penetralia of *The Building*, as the Pentagon was known, it was highly unusual for a one-on-one session to be requested by someone at that rarefied level of command. Normal protocols called for a variety of head-nodders from each intervening layer of the organization to be briefed on the topics and approve the presentation. On this day, that did not happen. At 12:30 P.M. I was sent in to the outer E-Ring office to meet General Stilwell. Actually, we had previously met when he was still on active duty and assigned as the Commander of U.S. Military Assistance Command, Thailand, and I was the Special Forces A-Team commander at a remote base called Nong Takoo. While I recognized him, he had no apparent recollection of our prior brief encounter, nor would I have expected him to.

Our discussion was quite pleasant and lasted about half an hour. We discussed many areas of phenomenology, but we probably focused on the Soviet efforts. One question he asked was where I was assigned and with whom I was working on these unusual topics. The meeting ended rather noncommittally and I left wondering what had just happened. It seemed that we had just had a pleasant conversation, addressed interesting topics that appeared to be of mutual interest, and thought that would be the end of it. Wrong again.

In reality, I was working concurrently with two different organizations, albeit one of them informally. The Inspector General projects I engaged in were important, but generally very mundane. The most intriguing project was the team to which I was then assigned that was exploring how the Headquarters of the Department of the Army Inspector General's Office could transform itself into something more useful. That study brought me into direct contact with our boss, Lieutenant General Richard Trefry, as I led an extensive interrogation into his understanding of current IG practices and his vision for transformation. After his retirement General Trefry later was recalled to become the Military Assistant to President George Bush, the Elder.

The other key person with whom I was involved was then Lieutenant General Maxwell R. Thurman, who at the time was the U.S. Army Deputy Chief of Staff of Personnel. Thurman would later get his fourth star and eventually commanded U.S. forces during *Just Cause,* the invasion of Panama that removed Manuel Noriega from power. Shortly thereafter, the sudden discovery of leukemia sent him into retirement. However, both men would remain as important figures in my life, even after we had all retired.

Returning to my desk, I dwelled little more about the unwonted meeting with General Stilwell. Then, at 4:20 P.M. that afternoon the DAIG executive officer walked in and said, "Tomorrow morning you don't work here anymore." As it turned out Stilwell had contacted both generals Trefry and Thurman and moved me under the latter. This abrupt transition marked the first time I was assigned outside the normal Army personnel system. In fact, I would never again be transferred by the official system, a point that both perplexed, and eventually angered, the personnel managers who always believed they knew what is best for all soldiers. Years later I met an officer whom I knew from a previous assignment in Hawaii. He mentioned that while he was assigned to the office that manages all U.S. Army Infantry officers, he had held my records locked in his desk door because there was no place in the system he could file it.

Moving Formally into the Psychic Realm

For a short time I worked directly for Lieutenant General Thurman, but my interests in Soviet exploitation of psychic phenomena soon took me to the dark side and the U.S. Army Intelligence and Security Command (INSCOM) where I first met Major General Albert "Bert" Stubblebine. A supporter of Task Force Delta, General Stubblebine also knew the Army had to make major adjustments and was willing to take steps that were both innovative and courageous. With piercing blue eyes, white hair, and a gripping handshake, Bert made an impression everywhere he went. He was a dead ringer for the actor Lee Marvin, and actually perpetuated the rumors that they were brothers. It is Stubblebine who was unfairly portrayed by Stephen Lang as General Hopgood in the movie version of *The Men Who Stare at Goats*. It was Hopgood who dashed headlong into a wall in an unsuccessful attempt at interpenetration of two material objects.

It was under the guidance and protection of General Stubblebine at INSCOM that I actively pursued a wide range of topics including various phenomena. Many of those projects are discussed in my first book, *The Warrior's Edge*, which was coauthored with Janet Morris. Those topics ranged from remote viewing and psychokinesis to firewalking, orgone energy weather modification, and primary perception. Experimentation in the last topic came under the tutelage of Cleve Backster, who is best known for discovering that plants communicate and are sensitive to their surroundings. The developer of the techniques used in modern polygraphy, it was Backster who first hooked up a philodendron and measured electrical output when the plant was stressed. Using oral leucocytes, we took the process much further and suggested applications for communicating with agents who had been abducted.

Of course, while I was assigned at INSCOM, UFOs from 1981 to 1984 were included in my mix of topics as I had wide latitude in what I wanted to investigate. In fact, when people would ask me what I did in the Army, I would somewhat jokingly respond, "I'm a freelance colonel." That was not far from the truth while I was working directly for Bert Stubblebine. However, as readers that follow closely remote viewing lore know, Stubblebine ran afoul of others in the Army senior leadership and retired rather abruptly. There can be a price to pay when creativity exceeds acceptable boundaries—even when it is successful.

Stubblebine's retirement left the newly appointed Lieutenant General William Odom in charge of all Army Intelligence. He wanted nothing to do with remote viewing or phenomenology in general or me in particular. They say you are known by your enemies, and I had made a powerful one in Odom. One of Stubblebine's last endeavors on active duty was to find a position for me outside of INSCOM. There was, it turned out, a three-star general who was Deputy Commander of the Army Materiel Command (AMC) who was favorably disposed to unusual topics.

That was Lieutenant General Robert Moore, and he was known to strike terror into the hearts of even the most experienced officers of any rank. Stubblebine and I visited him and described the problem in blunt detail. If left in INSCOM, I'd be crucified—institutionally speaking. General Moore agreed to take me in, and the next out-of-process personnel move was undertaken. While he knew of my interests in paranormal activities, I had to fill a position of some kind. Therefore, I was assigned as the only military officer working in an office comprised of senior civilians that managed the Army's technology base. Those were projects that were early in the research and development (R & D) phase and involved a broad look at emerging technologies. As a result of my assignments in INSCOM and AMC, my contacts in both worlds—intelligence and research—continued to expand near exponentially. Because of my R & D position, I was now contacting civilian research organizations and defense developers on a continuous basis. In addition, I maintained my involvement with Task Force Delta as long as it continued to exist.

Following his retirement Bert Stubblebine went to work for respected defense contractor BDM as the Vice President for Intelligence Operations. We remained in close contact for the next several years and that contact played a vital role in the development of the ATP project. The initial meetings would be held in the BDM secure facilities in Tysons Corner, Virginia, and located not far from my office.

In 1984, after only a few months at AMC, I was abruptly given an assignment to take over a huge, but failing project called *New Thrust*. The army had recently successfully fielded five new major systems including the M1 Abrams tank, the M2 Bradley infantry fighting vehicle, the Apache attack helicopter, the Patriot air defense missile system, and the Multiple Launch Rocket System (MLRS). The next advances, it was thought, would be technology sectors and it was the responsibility of New Thrust to coordinate these next generation weapons systems. The R & D projects included technology-based advances in precision-guided munitions, command, control, communications and intelligence systems (C^3I), reconnaissance, surveillance, and target acquisition systems (RSTA) and a few others. The breadth of this assignment would lead to a totally unanticipated promotion, but also provide a wide variety of technology contacts.

Introduction to the Skunk Works and Area 51

Among the people I met was Dr. Ron Blackburn, a retired U.S. Air Force lieutenant colonel who was then working at the Lockheed Skunk Works in Burbank, California. Tall, rail-thin, and bespectacled, Blackburn was as close to a functional paranoid as one can be. While I had evidence that the KGB previously had intercepted some of my calls, and acted on them, Blackburn just always assumed someone was listening. When meeting in hotel rooms he would compulsively unscrew the voice end of the handset and remove the components—even though there was no classified discussion ongoing.

It was Blackburn who first asked me if I had ever heard of the infamous Area 51. In the early 1980s, this facility was still not widely known inside the military, let alone the general public, even though it had been functional for decades. During this period, what was called at the F-117, or stealth fighter, was a very dark secret. In reality, it never was a fighter, but rather a small bomber developed to decapitate the Soviet nuclear strike capability. Stealth was not just an advance in technology; it was a war-winner—meaning a weapon system so sophisticated that it changed how we would defeat a Soviet attack in the case World War III ever materialized. Extreme security measures surrounded the development of this craft, and of course, that secrecy generated rumors. Those rumors ran from reports of alien spacecraft, to strange cadavers, and eventually live extraterrestrial visitors collaborating with the U.S. Government. In the annals of UFO mythology, Area 51 would assume a prominent, albeit unwarranted role. But make no doubt about it, strange things did fly in the desert—but humans designed and operated those strange things.

Among areas of common interest between Blackburn and me were UFOs. We discussed many possibilities related to who might be in charge of UFO research. We both thought that there was some organization, probably within the U.S. Air Force, which had the responsibility. But we acknowledged that whoever had the ball, there must be an interagency effort as well. Our assumption was that somebody must be in charge, and we were well aware of all of the prevailing stories and rumors. Roswell, we assumed, was a real UFO event.

The most likely situation regarding what happened to the crash material was what we called the *Raiders of the Lost Ark* scenario. That reference came from the final scene in that movie in which the enigmatic Ark of the Covenant was seen being transported into an extremely large warehouse, probably in Suitland, Maryland, and stashed away, then possibly forgotten. The assumption of the group was that a craft had been recovered and a team of top scientists called in to examine it. The technology was so advanced they had little success in understanding it. It would have been as if a stealth aircraft fell into a remote area of the world and was found by a tribe of primitive people. They would look at it, but without a basic understanding of the physics and engineering behind the development, it would be of little practical use to them. Our guess was that a team had looked at it, but was unable to make any advances. They would therefore figuratively bury the material, with the intent of revisiting it every few decades to determine if we humans had advanced technologically enough to understand how the craft operated and how to exploit it. At the heart of the matter would have been the propulsion system—and what must be a new energy source.

Still, we thought there had to be an organization that had oversight and responsibility for protecting the technology. Since we both held Top Secret-Sensitive Compartmented Information (TS-SCI) clearances, we wanted to be very careful so as to not jeopardize those coveted tickets as they were important to our livelihood. We also anticipated that if an exploration of the topic were conducted responsibly, one of two things would happen. The best scenario would be that we would be brought into the existing program, thus gaining the information we wanted. In addition, done properly, we believed that the ATP members could bring some new expertise to the table.

The less desirable outcome was that we would be generically advised of a project and then censured. In that process we would be advised to read certain material, then be required to sign papers stating we would never reveal what we had learned—all accompanied by severe legal penalties for failure to comply. During my career I did get caught up in what was called *"inadvertent disclosure."* In some ways, that can be the worst outcome, short of going to jail. In that process you are allowed to read basic documents about the project but generally devoid of significant details, and then told to never inquire about the topic again. That process can be an effective strategy for controlling inquisitive minds inside the system. Even given that possibility, we were willing to get our hands slapped. The upside would be that if we did not get the desired disclosure, we still would know that somebody was minding the UFO store.

The Players and the Rules

Willing to assume the risks, I set about forming the group. The initial desired outcome would be to determine who knew what about UFOs. First, I needed a Sensitive Compartmented Information Facility, known in intelligence circles as a SCIF. These are rooms that are built to detailed specifications to prevent anyone from eavesdropping via electronic surveillance techniques. There are no external portals and the entire facility is enmeshed in wires that prevent intrusion. To run this meeting I contacted Stubblebine, who could get us access to one of BDM's SCIFs. Out of courtesy, we did inform Joe Braddock, a renowned scientist and the B of BDM, that we were going to hold meetings there. He seemed mildly curious, but exhibited no strong interest one way or another.

To establish the ATP group we decided on certain rules. At that point in time the Freedom of Information Act, or FOIA, was in full swing and UFO inquiries proliferated. It was stated by government administrators tasked with responding that the congressional staffers who wrote the FOIA would never have guessed that so many UFO requests would be filed to the point where they actually clogged the declassification system. To legally avoid answering any UFO FOIA requests, I adopted the term *advanced theoretical physics*, assuming no one would make the connection and request ATP reports. Further, there were no written documents kept within U.S. Government agencies. One of the rules was that there were no written reports to be kept by anyone, though it now appears that the rule may have been violated by one or more of the defense contractors who participated. As the person responsible for conducting the sessions, I never wrote any reports before or after meetings.

Membership was highly restricted and it was literally an old boy network. The rules were:

> — By invitation only (nobody was allowed to come that we
>   didn't know)
> — Participant background and interest had to be demonstrated
> — Referrals could be made, but I reserved the right of acceptance
> — Minimum security clearance levels were TS-SCI at SI-TK,
>   no exceptions
> — As noted—there were to be no written documents about the
>   meetings
> — Participants had to cover their own costs

In Richard Dolan's recent book, *UFOs and the National Security State: The Cover-up Exposed, 1973–1991,* certain dates of the ATP meetings are listed. They are probably correct, and as I no longer have the list available I cannot discount them. However, his list of participants is only partially correct, and in some cases wrong. Since it was always the intent to keep attendance confidential, I will not reveal names except for those who have already chosen to be identified, or have authorized me to use their names. Membership included people from the Army, Navy, and Air Force, plus several from defense aerospace industries and some members from the Intelligence Community. Obviously, at least one representative from the Lockheed Skunk Works was present, and Dr. Bob Wood, then with McDonnell Douglas, has identified himself as a participant to other UFO researchers. While those dates of the formal meetings may be accurate for plenary sessions, my contacts and briefings took me to many, many more places than ever have been identified. Even I did not keep a complete list of locations and senior people that I briefed.

The first meeting held at the BDM vault resulted in considerable discussion about the nature of the UFO phenomena. It rapidly became clear that almost everyone present was familiar with the same rumors, but no one admitted to firsthand knowledge. At best, all of the information was second- or thirdhand if provenance could be traced at all. The basic assumption remained the same—there was some secret organization that had responsibility for the phenomena. However, everyone from an organization that seemed a likely candidate—either from the U.S. Government or aerospace industry—thought it was some other agency or group that was conducting the research.

That narrative remained constant throughout the life of the ATP and beyond. The importance of that finding cannot be overstated when it comes to understanding the role of various departments regarding UFOs. The key assumption across all agencies was *that somebody else was charged with responsibility for UFOs.* The ultimate answer appears to be that nobody does have that responsibility. While this notion runs counter to all of the conspiracy theorist's proclamations, that was the bottom line.

Interestingly, I was not the first person to independently make this observation. In reviewing Jacques Vallee's recent book, *Forbidden Science—Volume Two: Journals 1970–1979,* for the *Journal for Scientific Exploration,* it is noted that he had encountered similar responses several years before. Vallee, whom I count as a personal friend, had several decades of experience with UFO investigation and was an inveterate chronicler of events as they happened. During the 1970s he was involved with many of the same intelligence organizations that were participating in the ATP. After years of interaction, he describes his "reaction to the Intelligence Community: Some fascination, mixed with disgust." In giving numerous briefings, and holding discussions within the Intelligence Community, or IC as it is known in inner circles, Vallee also found the analysts' real interests focused on what other agencies were doing in the UFO field, not the phenomena.

Delving into the Intelligence Community Regarding UFOs

The IC is an encapsulated universe and knowledge is power. Even after intelligence shortfalls of 9/11 were widely reported, the IC, while it has improved on information sharing, still has a long way to go before intelligence integration is complete. It is also essential to understand that personal relationships are far more important than institution-to-institution associations. One purpose of the ATP was to form those personal bonds; an objective that was never fully met.

Over the first few meetings the discussions were quite general. Classified reports concerning UFOs were made available. Over the years I had access to both classified and unclassified material, often regarding the same incident. What civilian UFO researchers did not know was that about 99 percent of the material was actually in the public domain. The small part of the information that was not being discussed involved protecting sources and methods of obtaining the data. It is those sources and methods that are usually the most sensitive part of any report, as revelation of those techniques may cut off future acquisition of intelligence data.

As an example, there is an extremely well-documented September 1976 case involving Iranian Air Force fighters encountering a huge UFO north of the capital, Tehran. The case will be addressed in more detail later. Of note is that the secret classification initially assigned to the case had to do with the fact that U.S. intelligence personnel received some reports directly from the participants, and not through official channels. In short, we did not want the Shah of Iran to know we were getting information that he, or his agents, were not aware of though they later assisted in the investigation.

After a few meetings the ATP group had discussed enough evidence to understand the UFO issues were real, but that government agencies were not nearly as involved as everyone had expected. While some of the initial conclusions may sound like a blinding flash of the obvious, they are included just for the record. The observations included:

— We determined there was sufficient evidence supported by
     high-quality data to know that some UFO cases were real
     anomalies—not just poor observation or misidentification.

— There were cases involving military weapon systems that posed a significant threat and should be investigated.

— Multisensory data supported observations of physical craft that performed intelligent maneuvers that were far beyond any known human capability. Examples included:

- Extremely rapid acceleration (0 to 4000 MPH near instantaneously)
- Speeds that are not achievable by any manned craft today
- Very fast, sharp turns (90 degrees or more and producing g-forces that would exceed human survivability)
- Abrupt disappearance from radar (long before stealth technology was developed)
- Interrupting electrical systems without physical damage to them.

— There were cases that involved trace physical evidence.

— The Condon Report was a seriously flawed effort. (This was extremely important and will be discussed in detail in a later chapter.)

— Study of the UFO data could provide a potential for a leap in technology. This would not require access to a craft, but could be derived from scientific examination of the reports determining the theoretical physics required to achieve such results

Supported by that assessment, a rational argument could be made to support initiation of a formal new project researching UFOs. Issues included:

— Many new sensor systems were available, and new capabilities are constantly evolving.

— Cases continued to occur, and some with very significant implications for national security.

— Review of prior studies (Blue Book and Condon) proved them to be inadequate.

— It did not appear that this would be a duplicate effort—meaning the proverbial deep black program was in fact a myth.

— We had assembled a team that was capable of conducting appropriate research.

— That the advent of the Strategic Defense Initiative (SDI)
brought new requirements. Specifically, if we were
going to start a war with space-based sensors, it was
imperative that it would not be because of a misidentified UFO.

Based on the conclusions drawn from solid data, I developed a basic briefing and began working my way up the hieratical military system. This was an
information briefing designed to provide the recipient with sufficient knowledge to comment. Note that in the military, information briefings are very
different from decision briefings. In the latter case, the principal person is expected to make a specific decision such as committing resources or granting
permission for actions to occur. It is axiomatic that it is easier to get forgive-
ness than to obtain permission for risky ventures. As mentioned, I was prepared to get slapped if that was what it took to stir up the UFO pot. Mentioned
previously was inadvertent disclosure. I had experienced that personally. What
I learned was that you may not learn what is in the box, but you will certainly
know when you've stumbled into it, or stepped on the wrong toes. As I'll describe repeatedly, with one exception, I never encountered resistance. In fact,
quite the opposite almost always happened.

Taking UFOs to the Four-star Level

Washington leaks like a sieve! It is said that if two people in the Capitol know
something, it is no longer secret. Therefore, after holding three general sessions I decided I'd better let my boss, General Moore, in on the idea that I had
been holding secret meetings about UFOs. The last thing needed was for another *Washington Post* article to break without warning. Since such an article
would not be time-sensitive, it could be used for public titillation on any slow
news day. General Moore found the topic moderately interesting, but thought
that I should also brief his boss, General Richard Thompson, the four-star
commander of Army Materiel Command (AMC).

Unlike Lieutenant General Moore, who had a lot of research and development experience, General Thompson was primarily a logistician, and responsible for procuring and fielding all of the weapons systems for the Army—a
huge formal obligation. AMC was an organization of more than 100,000 personnel, most of whom were in the civilian workforce. While R & D was important, that funding paled compared to acquisition. General Thompson's job
entailed spending tens of billions of dollars to buy and maintain tanks, helicopters, missiles, sensors, communications equipment, and a myriad of other

things that supported the most advanced ground fighting force in the world. Having had extremely limited previous direct contact with him, I could barely imagine where a topic as bizarre as UFOs would fit on his priority list.

At that level, time is precious and information briefings short. In this case I was given fifteen minutes to tell him that I had been running meetings that, if exposed, would cause a lot of red faces—and probably burn my ass permanently. Read-ahead papers were always prepared so the commander would not be caught cold. In this case, it was very nebulous and stated I would address advanced theoretical physics—the letters UFO did not appear.

The session went far better than I had dared to anticipate. General Thompson had no apparent knowledge of the topic but was willing to assume the risk if a newspaper article appeared. Compared to the nasty public feud of the time over the efficacy of testing the Bradley Fighting Vehicle, UFOs would be noise. In that controversy, reporters made assertions that the automatic fire-extinguishing system would kill the troops inside the Bradley and that the results had been rigged to show the system was safe. By comparison, news of ATP meetings would be a minor blip. But, more important, General Thompson asked how he could help me. My response was that when the time was right, I would like him to introduce me to the four-star officers involved in R & D from the other services. He agreed and later he kept his word.

Who Was in the Loop

The first set of meetings and briefings were conducted from 1984 through 1988. What was not known publicly until now was that I was able to continue some of my efforts after retirement and while at Los Alamos National Laboratory (LANL). The presentation morphed over time based on reactions from the people I briefed and as new cases emerged. We actively sought out people that were thought could be helpful and provide connections to those who might have responsibility and interest in a UFO study. Note that a few of the team members attended some of the briefings, but I was the only person to have direct contact with all of the recipients.

Ben Rich and the Skunk Works

Fairly early in the process Ron Blackburn introduced me to Ben Rich, a brilliant aerospace engineer who had taken over as president of the Lockheed Skunk Works from the legendary Kelly Johnson. Johnson was the pioneer that chose Groom Lake and Area 51 for the development of the most advanced aircraft of their time. This was the initial home to the U-2 spy plane, which was actually reported as a UFO on numerous occasions. Seasoned pilots would spot something shiny above them and clearly under intelligent control. Based on incomplete knowledge of the real state of the art in aviation; they knew nothing could fly that high, thus they dutifully reported these objects. Of course, the U-2 community was happy to have their craft declared UFOs. In fact, that cover story remained officially classified secret for nearly half a century.

Ben Rich was involved in numerous advanced aircraft designs including the prototypes for the SR-71 and the F-117 Stealth Fighter. For exotic aircraft that pushed the envelope of engineering, Rich was the go-to guy for the CIA, the U.S. Air Force, and others. His experience and track record were unmatched. Therefore, we reasoned if anybody was in the know, it would be him. My several contacts with Rich spanned nearly a decade and continued after I retired from the Army. Of course our mutual interests covered far more than UFOs and included work on advanced aviation concepts for military purposes. I was the one who invited him to speak at Los Alamos and introduced him to Dr. Sig Hecker, who was the director of LANL in the early 1990s.

Rich was extremely attentive to what we presented to him about UFOs and even arranged to bring in a friend who once had been the Deputy Director for Research and Engineering (DDR&E) in OSD, and who was then a key member of the Defense Science Board. His friend was underwhelmed with our presentation and thought the evidence we presented was not sufficient. Short of showing him a UFO, I'm not sure what it would have taken to make a convincing case for him. Note that the DDR&E was the top research position in the Pentagon. The scientist was not at all evasive, just not convinced.

Rich, while remaining a bit skeptical, displayed significant interest. In fact, he had a shopping list of technologies that he wanted to get his hands on. The top priority was propulsion, but other technologies were of interest including navigation and the means for disappearing from radar. After all, it was Rich who had come up with American stealth technology. Begun as a DARPA project, Have Blue, Rich knew that if the locations of radar sites were known, then measures could be taken to redirect the incoming signals. Given the computing power available in the 1970s, the easiest calculations were based on straight edges; hence the aircraft design was akin to a number of flat plates. His approach ran counter to those guessing about what a stealth fighter might look like. That resulted in the Revell Toy Company airplane model they called the F-19, complete with flowing curves. While Rich's design decreased the radar cross-section from a known direction, it was increased in other directions. American stealth technology was not on par with the *Star Trek* Klingon cloaking device. Whatever UFOs were doing, the technology would have to have been different from ours and far more advanced. He wanted it.

Having heard the same rumors we had, Rich's first approach was to assign a key person on his staff to see if they could ferret out a black program on UFOs that was being conducted by some other organization. At the time the Lockheed Skunk Works had sensitive technology contracts with all of the usual suspects. If one of those organizations was involved, the crosswalk would be natural and germane to their projects. None was found. The person assigned, an experienced SR-71 navigator, became a liaison to me. The ATP group would benefit from his information but only one other participant knew his identity.

I am well aware of the quote attributed to Rich concerning having the technology to take ET home to the stars. The statement demonstrated that he, like many Americans, was frustrated with the propensity of government agencies to overclassify projects—a topic addressed in a separate chapter. The oft-repeated quote states, "We already have the means to travel among the stars, but these technologies are locked up in black projects and it would take an act of God to ever get them out to benefit humanity. . . . Anything you can imagine we already know how to do." Even if the quote is accurate, though it seems a bit of an overstatement of facts, we did have such capability at that time using nuclear propulsion. Rich did address UFOs in public, but was quick to point out he meant UnFunded Opportunities. After all of my first-hand interactions with Rich, I was, and remain, convinced that he did not have any direct knowledge of a surreptitious UFO program involving extra-terrestrial beings.

If Five People Know, Dr. Edward Teller Is One

Another key individual that rumors associate with knowledge of UFOs is Edward Teller. The rationale for the association was reasonable. While Dr. Teller described himself as the father of a son and daughter, he was known to the world as the father of the hydrogen bomb. At the time of the Roswell event in 1947 he was the leading authority in Western civilization in the knowledge of the most advanced energy source on the planet. Years later some scientists and political observers found some of his concepts quite radical, but at that time he was a man that President Truman was most likely to go to for answers to complex issues related to energy.

I first met Dr. Teller at Los Alamos. There was a briefing ongoing in a darkened room when I slipped into an empty chair at the table. When the lights came on, I was quite surprised to find that I was sitting next to him. Over a period of time I had several interactions with Dr. Teller and he became an early supporter of my work in nonlethal weapons. This progressed to the point that I had him to our home in Santa Fe, New Mexico, for dinner on a few occasions. Eventually, I did have direct discussions with him about UFOs.

For the first dinner on November 13, 1993, my friend, physicist Dr. Hal Puthoff, was also invited. The purpose was so that Hal could describe his groundbreaking theoretical research in zero point energy (ZPE) to Dr. Teller. The ZPE concept was a topic that Dr. Teller was not familiar with at the time, and this informal setting provided Hal a great opportunity to go one-on-one with him without playing to an audience. That was important as too frequently in technical briefings extraneous observers will interject comments just to prove how smart they are, or why a theory can't possibly work regardless of evidence. As Dr. Teller arrived at our home in Santa Fe, Hal turned to me and said, "My God, a living icon."

Even in his upper eighties, Dr. Teller's mind was extremely sharp and focused. He did not suffer fools, or trivial matters. One knew immediately if a pedestrian or inappropriate topic had been raised. With a thick Hungarian accent Dr. Teller would abruptly state, "That is not of interest to me!" Fortunately both ZPE and UFOs were acceptable topics for discussions.

At the dinner table Hal sat next to Dr. Teller and went through his presentation. Dr. Teller stopped him periodically and asked, "What is the reference to that?" Hal provided the reference, and Dr. Teller said, "Send it to me." Hal agreed and proceeded. Then Dr. Teller would again stop him, saying he understood the prior commentary, but what was the new reference? That process was repeated several times. Finally, Dr. Teller closed his eyes and remained silent for many minutes. At the time it seemed like an eternity and I was wondering if we needed to check his pulse. Then Dr. Gregg Canavan, one of his protégés who was also at the dinner, interjected finally, "Dr. Teller is thinking." After more than ten minutes he opened his eyes and stated, "That would mean the following," and he proceeded to tell Hal the implications of his theory. Hal noted that Dr. Teller was indeed correct. Later Hal confided in me that it took him a computer to do the sequential logarithmic equations that Dr. Teller had just done in his head. (You might note that I have referred to him as Dr. Teller for that is the degree of respect that was always shown to him in person. Even his protégés called him Dr. Teller.)

On another dinner occasion we discussed UFOs in general and several specific cases. My guess was that if the Roswell crash was real and only a very few people had been in the loop, Dr. Teller would have been one of them. Again, in 1947 he was working on the most powerful energy source known to mankind, and energy, not flying craft, would have been the dominant concern.

Interestingly, he did not appear to be familiar with the Roswell incident. As we discussed it, he came up with the identical hypothesis that I, and a number of people engaged in the ATP project, had derived. If a foreign crash had occurred, it would have scared the hell out of everyone. The first assumption would have been that the Soviets had made a major leap in technology. Given the state of the newly initiated Cold War, that would have caused panic in senior defense officials. If the craft were determined not to be of Soviet origin, then the logical place to take the artifacts would have been Los Alamos, not Wright Field in Ohio. The scientific capability at LANL would have exceeded that of the U.S. Air Force. As Dr. Teller noted, the national labs were established to handle the most difficult technical problems and he agreed that he would have been one of the people called to investigate—he wasn't.

One case we did discuss that caught his interest was Cash-Landrum, the December 1980 incident in which three people were irradiated. Again, he was not familiar with this event, but seemed intrigued by the facts surrounding the radiation injuries. I will cover Cash-Landrum in more detail later. That is a very solid case, in which the observations and facts just don't make sense or support any prosaic hypothesis.

Again, I'm well aware of the various statements pertaining to UFOs that have been attributed to Dr. Teller. They are at variance with my firsthand experience in directly addressing the topic with him. A few people aware of my relationship with Dr. Teller have suggested that he was being evasive, and just not admitting his knowledge of the topic. Based on my very careful observation of his demeanor during these conversations, I concluded that neither obfuscation nor deflection occurred. Rather, his reaction was as anticipated, and, more important, fits both the facts and common sense.

Burt Rutan: An Interested Skeptic

Another aerospace giant known for his high-tech innovation is also a personal friend. Burt Rutan has expressed his interest in UFOs with me and several other scientists who have rather detailed knowledge of the topic. Years before he formally rolled out *SpaceShipOne* at his hanger in Mojave, California, we had discussed space ventures in general and UFOs in particular on other occasions.

Unlike most aerospace developers and entrepreneurs who tell people what they plan to do, Burt tells you what he has done! On April 18, 2003, I was present when the *White Knight* rolled by and took off after Burt announced his intention to be the first civilian enterprise to put a man into space. Then slightly over a year later, on June 21, 2004, I was also honored to be on the tarmac as he successfully put Mike Melvill into space and fulfilled his promise. Standing next to me at that time was none other than Captain James T. Kirk of the USS *Enterprise* (William Shatner).

It was in October of that year that Burt claimed the ten-million-dollar Ansari X Prize for successfully launching a civilian craft into space twice within fourteen days—a monumental achievement, proving that space was no longer the sole domain of nations and that independent civilian industries were about to take over a key role in exploration beyond the bounds of Earth. The prize money was awarded to Burt's sponsor in the quest for the X Prize, Paul Allen of Microsoft fame, who had sufficient trust in the vision to invest far more money than the prize was worth. Allen did send some of the prize money to Burt's company, Scaled Composites, who, in their typically magnanimous manner, distributed it among all their hard-working and highly dedicated and motivated employees.

Burt's creative endeavors were acknowledged in *Time* magazine when in 2005 he was named as one of the one hundred most influential people in the world. Then in 2006, CBS's *60 Minutes* did a special segment about his adventures. Creative and renowned for his innovations, Burt is as conceptually involved as one can get in advanced aviation.

Just as I had experienced with others who are considered the insiders of aerospace, Burt is interested in UFOs but generally skeptical. He is, however, very suspicious, and even dismissive, of wild claims, especially those that suggest that a UFO has been reverse engineered. Burt's current exploits include a cooperative venture funded by Sir Richard Branson to develop *SpaceShipTwo* for Virgin Galactic that will take paying civilians into suborbital space. Again, these are the top innovators in the field, curious by nature, open-minded, and with unparalleled connections, yet they seem to have no indication of a secret UFO program.

NORAD—Threat or No Threat

While in the Army I conducted dozens of other briefings with senior military and Intelligence Community officials. With one notable exception the response was always the same—interest, skepticism, and no direct knowledge of a program that was responsible for researching UFOs. However, several of them thought there must be a program someplace else. The following are examples of some of the positions and responses I encountered.

One of the people I briefed was a general officer who was previously assigned as a senior watch commander in the North American Air Defense Command. Known to most people simply as NORAD, the massive underground facility is located in Cheyenne Mountain, where it was designed to withstand a nuclear blast. The organization was designed for the primary purpose of detecting an airborne attack from the Soviet Union. Our nuclear strategy at the time, mutually assured destruction, appropriately called MAD, required that a sneak attack be spotted and a retaliatory response be made immediately. This was accomplished by constant monitoring of an extensive set of radar systems that blanketed our northern skies.

When I inquired, the general told me he was unaware of any tracking of UFOs. While this ran counter to other information I had collected, it is consistent with his position in the organization. What I knew from other sources was that unidentified objects were spotted periodically. In fact, years before that, when on an Inspector General study at Fort Carson, Colorado, I took the time to visit to NORAD. While there I decided to take a chance and probe a bit. A young lieutenant was giving the unclassified public briefing and asked for questions. I asked, "Do you ever track objects that accelerate very quickly, or make extremely sharp turns?" Without blinking he responded, "You mean UFOs. Yes." He declined to comment any further. In fact, another U.S. Air Force officer who later participated in the ATP had provided me with unclassified data indicating that uncorrelated objects were spotted, probably once or twice a month. The discrepancy in information most likely suggests that the incidents were not of sufficient interest or concern to reach the flag officer level. Those who saw the reports, on the other hand, were personally involved and did take note.

The reason I state that the responses are consistent based on the position of the respondent is the difference in responsibility. While NORAD may have been state of the art when it was built, it became impossible, or at least impractical, to constantly upgrade their computers. The emphasis of the computer codes was to distinguish *threat* from *no threat*. Therefore, extraneous data were generally rejected, meaning if the incoming objects were not following a predicted path that indicated a threat, it was rejected. There would be no need to alert a senior watch commander. These differences are consistent with what is reported throughout the ATP project. Rarely did any UFO incidents take on real significance—they are more interesting to civilians who have trouble comprehending the Air Force's position on the subject.

The Defense Intelligence Agency Would Have to Know—Wouldn't They?

One of the most interesting interviews I conducted was with a retired three-star general officer who previously had commanded the Defense Intelligence Agency (DIA). The DIA is the totally military arm of the Intelligence Community and focuses their attention on national security threats to the armed forces. Their mission statement is to "Provide timely, objective, and cogent military intelligence to warfighters, defense planners, and defense and national security policymakers." And their vision is "Integration of highly skilled intelligence professionals with leading edge technology to *discover information and create knowledge that provides warning, identifies opportunities, and delivers overwhelming advantage to our warfighters*, defense planners, and defense and national security policymakers" (emphasis added). It seemed logical that the DIA would be an integral part of any UFO data collection program, or be involved at some level in understanding UFO capabilities as part of their vulnerability studies. After all, there were stories about how quickly military response teams arrived at UFO crash sites and other incidents. To accomplish that task, it would be through DIA that the networks were coordinated—if they ever existed.

On this occasion General Stubblebine made the introduction and accompanied me on the briefing. When I finished the lieutenant general noted that DIA did not have a requirement to collect UFO data. For those not familiar with intelligence gathering, everything collected is based on established requirements. The IC does not just collect interesting ideas. More important, however, the former DIA commander did go on to describe his own personal experience. Many years before, as a new lieutenant serving in Korea, he had observed flying craft making maneuvers that he believed would have been impossible for our Air Force at that time. The point is that his own belief system accepted the existence of UFOs based on his personal eyewitness observation, but that was not enough to generate collection and reporting requirements.

There was an interesting reaction from another very senior science official in DIA. After I concluded the presentation he said, "I always wondered why I wasn't briefed on this topic." The importance of such statements is that if an incident, like a UFO crash, did occur someplace in the world, it would most likely be an element of the IC that would obtain first knowledge and pass the report back to one of the headquarters mentioned. For recovery it would be imperative to have an apparatus in place that could respond. Even if it were a tightly compartmentalized project, you would still need to have senior leadership involved, if only to protect it or allocate the resources required to support the mission.

The Central Intelligence Agency Must Know

A sitting deputy director of the CIA gave General Stubblebine and me a similar institutional response. This was a person who was known to both of us, and had been aware of some of the more exotic programs. In fact a few years prior to this interview he had come to my home and observed metal bending firsthand. In our UFO discussion he stated that the CIA did not have any requirements for collection of UFO data. Of course there were some UFO reports in the CIA databases, and I'll later describe the impact of FOIA on their system. While the agency was not actively involved in UFO research, I did request a liaison from the CIA to the ATP. That was granted. He also provided me with other contacts both inside, and outside, his agency. We remained in communication after he left the CIA and he became an active supporter of my work in nonlethal weapons when I was at LANL. His civilian position after retiring from the CIA placed him near the apex of surveillance satellites. Although we maintained periodic contact he never mentioned having any additional information about UFOs.

The offer of assistance by one suspected of hiding a UFO project was just one of many examples in which help was given in my search for a black program. History has shown me that when you are deflecting a person from a black program, you don't provide them with assistance that could expose the project you want to protect. The responses given by this person were commensurate with those of other agencies. Again, there was no indication of deflection from the topic in order to mislead us.

My normal work brought me in contact with various elements of the CIA. While some of those visits involved discussion of phenomena such as remote viewing and psychokinesis, others were more mundane, including nonlethal weapons. Due to my reputation it was not uncommon to have topics like UFOs come up in conversation. Mostly people wanted to know what I thought about these things. But on several occasions officers would volunteer information about their personal observations. Surveys have shown that from 7 to 10 percent of the population have seen what they believed to be a UFO. Members of the CIA are no different. In no instance did anyone mention that the observation had been directed as part of their official duties. Rather, they reported being engaged in some normal activity and making the observation incidentally.

As many UFO buffs are aware, Dr. Bruce Maccabee, formerly an optical physicist for the U.S. Navy, has discussed his meetings about UFOs at the CIA. Bruce is one of the most skilled UFO researchers, a topic he pursued on his own initiative, and not related to his naval research. Over time he acquired a reputation as an expert in the field, primarily due to his in-depth investigation of intriguing cases, especially ones involving photographic analysis. We have discussed our encounters at the CIA and with other groups as well. Bruce also has found that after giving presentations on UFOs that people from these agencies would come up and share their personal observations. They also expressed an interest in what other agencies might be doing in the field.

A footnote to a CIA report by Gerald Haines, concerning that agency's involvement with UFOs, contained a September 30, 1993, memo by John Brennan to the Executive Assistant, Director of Central Intelligence, Richard Warshaw, that mentioned a response team. Brennan stated, "The CIA reportedly is also a member of an Incident Response Team to investigate UFO landings, if one should occur. *This team has never met*" (emphasis added). The poorly written footnote has been grasped upon by true believers as proof of government interest. While creation of a crisis response team seems logical for a host of contingencies, the apparent lack of interest reflects the priority that the CIA gave this matter.

The National Security Agency—They Listen to Everything

Briefings at senior levels of the National Security Agency (NSA) produced similar results. Howell McConnell had been participating in the ATP from the beginning. McConnell was a career employee at No Such Agency, as some people jokingly refer to this supersecret organization, which was located a stone's throw outside the Washington Beltway at Fort Meade, Maryland. Based on his own initiative and personal concern about UFOs, Howell had written a position paper on the topic that was made available to his superiors. Upon a document search responding to a written request under FOIA, that paper had been released by NSA. Thus, he was known to the public as informed in the area. For personal reasons Howell and Jack, a coworker of his, had a deep interest in the topic. At some briefings Jack described his own sighting that defied explanation. He reported seeing an object moving across the dark sky with a stream of light trailing behind it. According to him, the object abruptly stopped and the light seemed to be sucked into the source, much like a string of spaghetti might be vacuumed into one's mouth. The light did not go out, but rather it seemed to be withdrawn into the UFO; something that lights don't do.

When it came to recordkeeping, Howell was also a saving grace in NSA. While civilians seem to believe that every piece of information that goes into the U.S. Government is saved, that is not true. Especially in the days of paper files, Washington, D.C., would have sunk into the swamp on which it is built just under the sheer weight of the documents. Rather, most material was scheduled for routine destruction after a specified date. Using an old boy net, McConnell had become known inside NSA for his interest in UFOs and incoming reports were regularly funneled to him.

He kept many of those reports long after their expiration date. It is important to note that the vast majority of those documents were from intelligence officers around the world who were sending in reports that appeared in their local newspapers. There were a few that got our attention, as they came from trusted sources. However, even having reports with high credibility, and possible military significance, was not sufficient to get formal collection requirements related to UFOs.

As ATP progressed, Howell, Jack, and I did meet with a very senior official in NSA regarding what had been concluded thus far in our study. Like the senior officials from the other agencies, he was interested but had nothing to offer. There was no problem with Howell keeping track of information regarding UFOs, but no requirement to do so. Most of the UFO reports that came into NSA were generated by people located at foreign sites passing on routine information they thought worth observing.

Another set of important briefings were in Air Force channels. As it turned out, I was requested to give a briefing at Kelly Air Force Base in San Antonio, Texas. The subject was not UFOs, but on the process that AMC was using to transition basic technology into engineering development. By comparison, the Air Force projects to develop new aircraft dwarfed the Army R & D programs as they seemed to be very adept at convincing Congress to provide them lots of money. This was a high-level meeting with the four-star officer involved in R & D of each service present. I took advantage of the situation and asked General Thompson to mention me to General Larry Skantz who then ran U.S. Air Force research and development and was headquartered at Andrews Air Force Base, just south of the Washington Beltway. General Thompson agreed, and during a break we made a brief handshake introduction. That was all that was necessary to gain concurrence to allow me an audience at a later date.

A few weeks later the meeting was held. Due to a bit of nervousness in the Army Headquarters in the Pentagon, Dr. Jay Scully, the Assistant Secretary of the Army for Research, Development, and Acquisition, insisted that someone from his office accompany me as a head-nodder. Actually Scully had already heard my UFO briefing and was generally supportive. However, prior to that occasion I did not enjoy a cordial introduction to him. It was not regarding UFOs, but rather the process for technology transition that went astray. Then a lieutenant colonel, I arrived for the meeting alone, which was a huge mistake. Protocol dictated that when meeting someone holding an assistant secretary position, a general officer-level person would accompany any staff officer to the briefing. My senior executive staff (SES) boss failed to make the meeting and I didn't know where he was. Worse, an article had appeared in *Defense News* the day before stating that Dr. Scully was on the outs in the Pentagon and that no one took him seriously anymore. Without my SES, I was proving this insult to be true. We sat at the table, he asked where my boss was, and then literally picked up my briefing and threw it at me. Not a propitious start. We did kiss and make up later with Lieutenant General Moore apologizing. That R & D transition process was then approved.

The UFO briefing went better, but Dr. Scully wanted his own man present in case things did not sit well with General Skantz. We entered the office accompanied by the Air Force officer who had been working with ATP and a brigadier general who was in charge of advanced R & D. While the meeting was generally cordial, General Skantz asked why there weren't more blue uniforms (meaning USAF personnel) in the room. Both generals seemed interested, but obviously had no real knowledge of the topic. In the end I was directed to go to the USAF Space Command, which was then a two-star command located in El Segundo, California, near LAX, and get their reaction.

That meeting was set, and the U.S. Air Force officer from ATP and I made the trip. He was the same officer who had provided me the data from NORAD and had the technical expertise to explain it. The commanding officer was then Major General Don Katina. The meeting went well and we asked for support establishing a more formal program. Since General Skantz had directed the visit, we needed a specific reply that we could send back to him. After a lengthy discussion, he told us, "Tell him Katina says yes." An important point about this meeting was that, while open-minded, he did not have any apparent background in the UFO subject. In a few years Space Command would become a four-star command at Peterson Air Force Base in Colorado, and General Katina would get his fourth star and take command of that organization as well.

The Strategic Defense Initiative—Going for the Gold

After setting the stage with the services, it was time to transition the project into a formal program with a real budget. The logical place to go was Strategic Defense Initiative, known to most people simply as SDI or by the colloquial name—Star Wars. President Reagan was already concerned about ballistic nuclear weapons even before he was elected. Pushed by him, SDI had become the most massive and controversial program in the Department of Defense. Headed by Lieutenant General Jim Abrahamson, SDI had a budget of about five billion dollars. (That would be the equivalent of about ten billion dollars in 2010.) We figured that SDI certainly had both the money and the mission that could support a small program to study UFOs.

For several years we had been postulating that SDI would be the appropriate place to ask for funding. When completed that project would monitor space for potential incoming missiles. As decision and response times decreased, it would be essential that the command and control system have accurate information regarding objects coming at the United States. Our concern was that an uncorrelated target—meaning a UFO—might trigger a response based on erroneous data. In short, could a UFO accidentally set off the next world war? However long the odds, it seemed that having knowledge of everything that might be flying in critical airspace would be prudent.

Again it was General Stubblebine who was able to arrange a meeting. General Abe, as he was known to his staff, was not aware of the topic for discussion and his protocol people did not demand the traditional read-ahead paper. For this meeting I had more help, including Howell McConnell and Jack from NSA as well as the USAF staff person who had worked with me. Howell was important as he brought along some very sensitive documents that were actually intercepts of Soviet discussions about UFOs.

I began by introducing everyone; then tried to easily transition into the UFO matter. Within a couple of minutes, General Abe stopped the meeting. "Who are you guys really?" he asked. He seemed a bit stunned by the topic. This was not deflection, but a clear lack of prior knowledge of the topic. Obligingly we again went around the table introducing each person and the agency he represented. While the Air Force data from Space Command got some interest, it was the NSA material that received the most attention.

What the Soviets Were Saying

Included in the briefing were statements from Dr. F. Yu Zigel, a senior Soviet astronomer and laboratory director. Zigel's comments included:

— "The sightings demonstrated indisputably artificiality, strangeness, and intelligence."

— "Unusual speed and kinematic movements, luminescence, invulnerability, and paralysis of aggressive intentions" (inferring the Soviets had tried to attack one or more UFOs).

— "To explain these events by natural causes is senseless."

— And, "the only hypothesis that offers an adequate explanation is UFOs."

Obviously, the Soviet scientists that NSA was listening to did not doubt the reality of the UFOs.

"How did you get this?" General Abrahamson asked, seemingly shocked by the verbatim transcript. McConnell responded, explaining our technical eavesdropping methodologies. He then asked Howell, "Can they (the Soviets) do this as well?" General Abrahamson was asking about Soviet signal intelligence intercept capability. "We think they can," responded McConnell. What shocked me most about the meeting was that the senior officer running the most expensive and highly secretive program in the world did not seem to understand the threat capabilities. That was more surprising than the UFO discussion that followed.

As we moved ahead with the meeting and presented more data, the tone changed considerably. General Abrahamson noted that as a former fighter pilot the concepts intrigued him. However, when it came to money, he turned us down. General Abrahamson noted that he was "doing some hairy things," but if he got caught funding something as exotic as a UFO project, Congress would use it as a weapon against him in the perennial budget wars on the Hill. The Star Wars program already was extremely controversial and his adversaries, of whom there were many, would conclude that if he could afford to fund a UFO project, SDI must have too much money. This was an entirely reasonable position as there was no doubt that many in Congress had their knives out for this program. His budget already was under serious attack and he did end up taking a billion-dollar cut that year. Despite wanting money to formalize the project, I had to fully agree with his rationale. It was also abundantly clear that Star Wars was in no way related to fighting ET—that was not even on their scope!

Anybody who has participated in the U.S. Government budget process understands just how vicious the infighting can be. Despite prior buildups, in the late 1980s the annual defense budgets were already declining. It was a zero-sum game. If you wanted to fund a new project you had to figure out what other project you were willing to cut. Contrary to the protestations of the true believers, UFOs were nowhere on the DoD priority list. More on that subject later.

General Abrahamson was not totally dismissive. He told us that his program would soon be monitoring more of space than had ever been done before. If we could tell him what to look for, he would consider including that data in the algorithms that were being generated for the SDI system. For reasons not related to SDI, we did not have the opportunity to follow up on his offer.

Aside from the ATP effort I had more firsthand evidence confirming that SDI was certainly not designed to fight aliens from outer space, nor did they even have a basic knowledge of the UFO topic. In addition to my onetime discussion with General Abrahamson, I became close personal friends with one of his most senior people. As a top-level scientist, this person had in-depth knowledge of the capabilities of the entire U.S. space architecture, including, but not limited to, SDI. A true skeptic, he was open to the UFO topic, but stated he has never seen any data that supported the existence of a deep black UFO program—or that any organization was routinely keeping track of sightings or data related to UFOs. Over more than a decade and a half we have discussed many sensitive topics. There have been times when he has told me that specific topics were off limits and could not be discussed. UFOs were not one of them.

Only once, in all of the dozens of briefings that I gave, did I get a negative response. In all other situations general interest was shown, along with some reasonable skepticism stated. However over time there was more and more evidence that no formal UFO program currently existed. On this one occasion, things turned ugly. After I briefed General Max Thurman, who at the time was the Vice Chief of Staff of the Army (the number two four-star position in the Army), he asked me to address the Board on Army Science and Technology (BAST) study group that was examining the Army use of space. Heading that committee was Dr. Walt LaBerge, an eminent scientist who had previously served as the Principal Under Secretary of Defense for Research and Engineering, the Under Secretary of the Army, the Assistant Secretary of the Air Force, and was then a vice president at Lockheed. Among his many accomplishments LaBerge was the co-inventor of the Sidewinder air-to-air missile, and he had led the team that built the NASA Mission Control Center. His distinguished career had brought him in close contact with many advanced developments in the U.S. space programs.

Attending the meeting were about ten scientists and Dr. LaBerge. For about an hour I ran through my briefing and informed them that the universe was probably not built the way they thought it was and UFOs deserved serious consideration. As I concluded, Le Berge asked the group if they had any questions. There were a few pretty good questions from the attendees who seemed interested in what I had been saying. Then, the most amazing thing I have ever witnessed in a scientific meeting took place. Livid, LaBerge slammed his hand on the table and began screaming at me. His first comment was, *"You're not supposed to know that. That's what you learn when you die!"* To say I was startled by his overreaction would be an understatement. Never would I have anticipated that someone of his stature would totally lose control of his emotions in front of a group of peers. There may be outbursts when scientists disagree about facts, but bringing religious beliefs in an uncontrolled and explosive manner before a panel like this was far beyond the pale.

His outburst did not stop there, and LaBerge went on ranting for several minutes. He brought up another program, one that we were the only two people in the room who understood what he was railing against. It was the remote viewing program that he had been briefed on when he was serving in the Office of the Secretary of Defense. In 1987, Star Gate was still in the black world, and very few people knew about it. While he obviously had a strong emotional bias against the program, I knew from other sources that he was lying about the facts. As I picked up my viewgraphs I muttered to myself that I thought we were having a scientific discussion.

The Army Science Board—a Positive Response at Last

While there were many more briefings and discussions with officials who would likely have been in the loop, I'll conclude this chapter with just one of them. At the request of Dr. Scully, who was in charge of Army R & D, a meeting was established for a small group of members of the Army Science Board (ASB). Most of the attendees were not aware of the topic until they arrived at the Pentagon. This was probably the longest briefing conducted on UFOs in the entire ATP experience and it lasted most of a working day.

I opened the session with an expanded briefing that lasted nearly two hours. Pragmatically oriented, I focused the group on two issues: defense-related cases, and aircraft safety incidents. Of course the Space Command data and NSA materials were included, and presented by representatives from those agencies. Among the experts present was Richard Haines who was then working at NASA–Ames in California where he served as Chief of the Space Human Factors Office. With decades of experience in both psychology and UFO studies, Dick gave an excellent briefing on the results of his extensive research into reports provided by predominantly civilian pilots. At that time he had a database of more than 3,000 cases, some of them pointing to a clear danger to civil aviation. Those incidents have continued and many more have been reported over the past two decades.

In addition U.S. Air Force Major General Jim Pfautz gave a presentation from the point of view of his prior position on the Intelligence Community Staff. Stubblebine and I had met with Pfautz a couple of years prior. At that point in time, the IC was not organized the way it is today. This was long before the creation of the Director of National Intelligence, and the head of the CIA actually wore two hats. While he was the CIA Director, he was also the Director of Central Intelligence or DCI, a separate position altogether. Under the DCI there was an organization that was responsible for integrating intelligence data from all of the agencies and establishing a single IC position on any given topic. When the President was briefed, it had to be a unitary position that was provided, even if dissenting views were noted. The DCI could not brief the President stating that the CIA thinks one thing, but DIA or NSA has other opinions. Therefore, it was the IC staff that had oversight of almost all intelligence matters. General Stubblebine and I believed that if there were a program or networks that followed UFO data, it would be this group and the IC staff that would have to be in the loop. Previously we had met with Pfautz at his office and he had told us they were not the integrating agency for UFOs. That is important as his position was the place where reports came together. If there had been one office that would catch reports from any agency, it would have been under Pfautz. Also, if he were attempting to hide a program he would not have appeared at the ASB meeting.

His presentation was fairly general, bounced lightly off of the UFO topic, and then he alluded to the remote viewing program, without addressing it directly. He did seem to support the creation of a project that might gather and analyze UFO reports. Interestingly, he decided to expose himself to author Howard Blum, when *Out There* was being written. While it has been written that I'm at least part of the model for Blum's character named Colonel Harold E. Phillips, I can state that most of the book is pure poppycock.

At the end of the meeting, the members of the Army Science Board present discussed what they had heard. My pitch had been for the creation of a small but formal project to explore the topic. The members present unanimously concluded that there was sufficient evidence of high-quality observations and data from veridical sources to proceed. They pointed out that there was a need to be able to accurately assess space threats. There was one area on which they were split—that was whether or not the Army should be leading the project.

The majority agreed that the Army should take the lead, at least initially. Any such innovative project needs a champion who would be able to lead the charge into what would surely be a contested area. The minority vote was that it should go to the U.S. Air Force. While they had been somewhat interested observers, there was no apparent burning desire to get more actively involved.

The project leadership issue turned out to be moot. Returning to my desk at the Army Laboratory Command in Adelphi, Maryland, I found a note to call Colonel Branch at the Army personnel office in the Military Personnel Center. After a decade under their radar they had finally caught up with me. An assignment was offered that only required any full colonel who could breathe. They were adamant that I would not be allowed to find a new position on my own. The decision was fairly simple. I was then currently involved in a number of very advanced high-technology programs, ones that would likely be attractive to defense contractors, while the new assignment was a dead-end job that would blunt my expertise. What I did not know is that I would turn down a major defense contractor and take a position at Los Alamos National Laboratory (LANL). For the record, I was able to hold a meeting similar to ATP at the lab. Like the initial ATP that I orchestrated for years, there were no written records kept.

Tom Clancy's Commentary—More Accurate than Most Might Think

Those people who have read *Future War: Non-Lethal Weapons in Twenty-First-Century Warfare* know that the foreword was written by my friend, Tom Clancy. Tom had included some of my nonlethal weapons concepts in his books and in so doing allowed them to be introduced to far more people than I could ever have reached. Over the intervening years we have had intense discussions on a wide variety of topics. When I first broached UFOs and told him about the ATP he was quite skeptical. His thinking has come a long way since then.

People who have followed Tom's nonfiction works know that he has many very high-level connections in the military and Intelligence Community. Also, his fiction books are replete with technical details of weapons and sensor systems. They are not always quite accurate, but then he has intentionally made many changes to protect some of the classified aspects of our military capabilities. Obviously he has a cadre of people like me who feed him information—some of it quite sensitive. I remember one time visiting his home on the Chesapeake Bay and learning that a day or two prior Colin Powell had been there as well. That is an example of the people who stop by. During one of our numerous UFO discussions we talked about the rumors of a black program based on the Roswell crash. Tom's comment was quite informative. He said he knew we did not have a craft "because somebody would have told me!" That is probably an accurate statement.

Summary

This chapter is important as it describes how over several years my exploits took me to the senior levels of the military, and included briefing the director or deputy director of the CIA, DIA, NSA, plus many other officials in various agencies, industries, and government science boards. The outcome of such meetings almost always included being introduced to other influential people. Among those I met with were Dr. Edward Teller (who became a personal friend while I was at Los Alamos), Ben Rich (legendary president of the Lockheed Skunk Works who developed the SR-71 and the first stealth aircraft), Burt Rutan (aerospace engineer extraordinaire), and Lieutenant General Jim Abrahamson (then director of the five-billion-dollar Strategic Defense Initiative (SDI)).

What I learned in my personal, face-to-face meetings runs counter to the wild tales that abound in conspiracy theory literature. However, unlike those conspiracy theory proponents, I am the only person who has discussed this sensitive topic with each of those listed, and many, many others like them. Worth noting is that even at the most senior levels, some officials shared the conventional mythology of the general population regarding UFOs. Despite their high-level positions, they all seemed to think that someone else was responsible for the topic.

In studying UFOs, nothing is simple. The next chapter brings the conundrums that arose when a respected Army officer made audacious claims about crashed UFO material and the role he stated it made in advancing American technology.

2

THE CORSO CONUNDRUM

Curiouser and curiouser!
—LEWIS CARROLL, *ALICE'S ADVENTURES*
IN WONDERLAND

Of all the books concerning UFOs, *The Day After Roswell,* by retired U.S. Army Lieutenant Colonel Phillip Corso, was the best seller. It was, without a doubt, the single book that had the most impact on the field in recent decades. The book was purported to be the quintessential insider's revelation of information that the U.S. Government did retrieve a UFO at Roswell, New Mexico. Corso also confirmed that alien bodies were found and provided unprecedented information about what they termed extrabiological entities or EBEs. Further, it was stated in the book that the unique material recovered at the crash site was used to assist American technological advances. Then, on the dark side, Corso claimed that there was actually an ongoing war between Earthlings and the alien invaders. Highly successful, when released contiguous to the fiftieth anniversary of the famous Roswell incident, the book sold more than 250,000 hardcover copies, and reignited interest in this most intriguing mystery. This was, after all, one of the few first-person accounts to come out in many years. A former government official was admitting what many people around the world suspected—there was indeed a cover-up by the U.S. Government. Unfortunately, few of the extraordinary claims had any truth to them.

There is far more to this story than is generally known. For the record I must state that Phil Corso was a personal friend. Even though we disagreed about many of his statements I tried unsuccessfully to help him get a movie deal. In addition, I was with Phil at his home in Florida between his two heart attacks that occurred less than a month apart. That visit was on July 2, 1998, which was just two weeks before his death. A few UFO enthusiasts who have heard my comments have accused me of speaking ill of the dead as Corso cannot defend himself. While many of my observations may seem devastating, there is nothing included here that I didn't say directly to him. In fact, a copy of a letter I sent to him following the publication of his book is located in *Appendix A*. There is one fact that is generally agreed upon regarding Phil Corso: almost everybody that knew him, including me, thought he was a consummate gentleman and basically a great guy.

Discovering Corso

Thanks to George Knapp, a seasoned Las Vegas television investigative reporter, Phil Corso came to the attention of those of us at the National Institute for Discovery Science (NIDS) long before his book was published. NIDS had been established by a local real estate developer, Robert Bigelow, to examine scientifically two specific anomalous areas. One was the continuation of consciousness beyond physical death, and the other was UFOs. George Knapp had a long-standing interest in the area of UFOs, and at one point relinquished a coveted anchor position on the local CBS affiliate, KLAS-TV, in order to pursue a national television documentary on UFOs. Since then he has returned to KLAS-TV and, in 2008, won a highly prized Peabody Award for exposing nefarious relationships between water, power, and politics in Nevada. Knapp is a popular and highly regarded part-time weekend host on the late-night radio program *Coast to Coast AM*. He is also well known to UFO enthusiasts for his role in questioning activities at the infamous Area 51. In 1989 George made headlines in the UFO community by bringing forth the contentious story of Bob Lazar and his mythical experiences in the desert into the public domain.

What is not known by the public is that years prior to the publication of *The Day After Roswell*, it was George Knapp who had an agreement with Corso to be the one to pull the book together. Unfortunately for Knapp, that deal fell through and Bill Burns became Corso's coauthor. As a veteran investigative reporter, Knapp acquires information from many confidential sources for a reason shared by few reporters—he is trusted. This was the case with Corso, and George had kept the secret until it became clear that assistance in getting the book out was necessary.

While we were skeptical of the basic story, in March 1996 George made arrangements with Corso to have a small group visit him. With Corso's permission, Hal Puthoff, George Knapp, and I traveled to Stuart, Florida, and spent a couple of days listening to him describe his extraordinary experiences. Arriving one evening, we drove to the home of Phil Corso, Jr., and met with his father. At first, there was considerable difficulty in getting a complete story as young Corso seemed extremely skeptical of us, and kept insisting that his father not divulge the details of his exploits. It took most of the next day, but

Phil, Sr., opened up and told us about the basic incidents that are described in the book. Most striking to me was the consistency of his detailed information. That would continue to be true over the next few years. It was almost as if the speech was recorded. At any given time he would begin talking about specific events, and the details never changed.

The first meeting was sufficient to get our collective attention. In April 1996 Bigelow extended an invitation for Corso to come to Las Vegas and discuss the issues in more detail. Corso agreed to come for three days. To support this debriefing, both Hal Puthoff and well-known UFO investigator Jacques Vallee flew in to join us. Corso's story never wavered, but contained both plausible and troubling information. Plausibility came from his reported assignments supported by documentation. Troubling were topics that ran counter to the known history of technological developments.

This was not Corso's first major exposé of sensitive material. In November 1992, Corso had appeared before a Congressional panel investigating the fate of American servicemembers declared missing in action (MIAs) and prisoners of war (POWs). At the time he described a shooting war that had taken place between the former Soviet Union and the United States for a number of years. Except for a limited number of people in the Intelligence Community, almost no one knew about these incidents. While the world had heard about the downing of Gary Powers in his U-2 spy plane, very few were aware that over the past decades there had been numerous other events in which our planes were shot down. Even the families of the crews involved were told that their loved ones had died in training accidents. Many of the details were very ugly, including the fact that some crew members had been captured alive, exploited, and then killed. Unlike spies who were caught and traded on a bridge in the middle of the night, officially these aircrews didn't exist, so therefore an exchange was impossible.

Corso also reported on the extremely sensitive issue of POWs still missing in North Korea. He had served as the head of the Special Projects Branch/Intelligence Division/Far East Command under the tenures of three highly esteemed generals, General Douglas MacArthur, General Matthew Ridgeway, and General Mark Clark. To military personnel, what he stated was at least as shocking as the presence of aliens. Corso presented testimony, which was supported by others, that nearly 900 American POWs had been left behind in North Korea after the exchange took place. He stated that about 300 sick prisoners had been within 10 miles of the repatriation site at Panmunjom, yet never made it to the site, and were never accounted for. For political expediency, and counter to our most precious American values, the U.S. Government forsook its own soldiers and left them condemned to die in North Korea and elsewhere.

Following our meeting in Las Vegas, I spent a week in Washington checking Phil's background and included a trip to the Army War College, located at Carlisle Barracks, Pennsylvania. It was clear from our discussions that Corso had indeed had many high-level connections while on active duty and beyond. He was credited with being a military advisor to President Eisenhower, and later served supporting the Warren Commission as it investigated the assassination of John F. Kennedy.

Among Corso's closest allies was Lieutenant General Arthur Trudeau. It must be said that they had an extraordinary relationship given their rank differential. Lieutenant General Trudeau was a legendary visionary for the Army in research and development. In the early 1970s the Army initiated an oral history project that included only eight former generals. Trudeau was one of those selected for a series of in-depth interviews that would continue over the span of a year. When those interviews took place, some ten years after the retirement of both officers, Corso was found in Trudeau's kitchen. As part of Trudeau's history, Corso also was formally interviewed as part of the project.

However, when it comes to Corso's past, little is easy to follow. The basic reported path seemed accurate, with a fair amount of fluff involved. For example, at the end of World War II, as a captain he had been assigned to Rome. As he stated, he did have contacts at the Vatican. He appears to have been involved in obtaining passage for 10,000 Jews to Palestine. But then his memoirs state that he "was handed the responsibility for intelligence and security of Rome." He then goes on to claim that he personally "restored law and order" to a city he said was in chaos. However, when reading the awards he received for his service in Italy, they are commensurate with what would be expected of an Army captain doing an excellent job, but nothing as truly extraordinary as the claims. Obviously, and as would be expected, there were more senior officers that held the responsibilities described in his writings.

Discrepancies Quickly Emerged

There is an office in the Pentagon that holds all of the phone books that ever existed there. Amazingly, I found a woman who was the longest serving civilian in the Pentagon; she had arrived in 1942 before the building had been completed and was still there in 1996. She had a remarkable memory and actually remembered both Lieutenant Colonel Corso and Lieutenant General Trudeau. Considering the tens of thousands of midlevel officers who had been assigned to the Pentagon during her employment, accurately remembering two of them was quite a feat.

She quickly located both a phone book and an organizational chart for the Army Headquarters at that time in question. What was found only partially supported Corso's claims concerning his assignment. Corso stated that he had been transferred back to the Pentagon under Trudeau, and a division had been created for him called Foreign Technology Division (FTD). This is not to be confused with the large Foreign Technology Division of the U.S. Air Force at Wright-Patterson, Ohio. It was learned that the Army FTD was formed as Corso stated, and then disappeared shortly after his retirement. However, while Lieutenant Colonel Corso is listed as assigned to that office, he was not named as the director, as he had claimed. Rather, it was a Colonel T. H. Spengler who was listed as the head of the FTD in room 2D343 of the Pentagon. On several occasions, and in his writing, Corso specifically claimed to be the director of FTD when multiple documents did not support that assertion. When Corso was asked about the incongruence, he could only say he did not know Colonel Spengler. One of many conundrums.

The investigation went further. Corso recommended that we talk with General George Sammet who then lived in Orlando, Florida. At this juncture we brought in Lieutenant General Gordon Sumner with whom I had worked while at Los Alamos. General Sumner had been fired by President Jimmy Carter when he testified before Congress that the plan to relinquish the Panama Canal on short order was a bad one. After President Carter left office, President Ronald Reagan recalled General Sumner and appointed him as a Special Assistant to the Secretary of State for Latin American Affairs with the personal rank of ambassador. Having traveled with Gordon in Panama, I knew he was revered by many Panamanians and had a great breadth of knowledge about the area. Importantly, he was open to my discussions of unusual topics, knew General Sammet, and was willing to go with me to make the introduction.

General Sammet had a unique background and had held several positions from which he had direct knowledge of the topics addressed by my questions. When Corso was at FTD, then Colonel Sammet had been Lieutenant General Trudeau's chief of staff—meaning he would have had access to most of the information about major projects being conducted in that office. Although we had not specified the topic for discussion for our meeting he came prepared with a packet of information about Corso. He confirmed that Corso and Trudeau often had long, intense conversations about strange topics. However, when it came to UFO material retrieved from Roswell, he stated he had no direct knowledge of the topic. Interestingly, Sammet continued his career and eventually took over as chief of the Office of Research and Development, the position previously held by Trudeau. Therefore he was asked if in that position he ever encountered projects related to the Roswell material or UFOs. General Sammet again stated he had not.

This lack of programmatic continuity would have been very strange. Programs and projects normally are passed down from one chief to the next who assumes that position. If there were projects as important to technological developments as Corso claimed, certainly there would be a record of them. The logical person to have inherited the UFO program, if it existed, would have been General Sammet, and yet he denied ever knowing about it. This apparent discrepancy was pointed out to him with the comment, "Therefore we shouldn't believe Corso?" He responded, "I didn't say that." Institutionally, this lack of continuity for a significant research project does not make sense. The responsibility for monitoring progress should have been constant.

Errors Abound—Both Minor and Egregious

When *The Day After Roswell* was published I sent Corso a seven-page letter addressing the errors that I found in the book. Again, that letter can be found in Appendix A. My observations ran from simple mistakes, such as it is Adelphi, not Adelphia, Maryland, to glaring errors including the assertion that the Cold War was a cover for fighting ET; that the United States and Soviet Union were always cooperating in that war; and that we had an established relationship with ET. In my view, these statements are preposterous.

At the time of our early meetings, Corso arranged for us to be provided with a copy of his original manuscript. From a quick read, it was obvious that he desperately needed a coauthor, as much of the material rambled on, sometimes almost incoherently. Also, Phil was not computer-literate. There were pages and pages of handwritten material, some of which had been later typed, and even a few drawings. Importantly, there are very significant differences between the original handwritten manuscript, and the published version of the book. The discrepancies start at the beginning of the book. Phil's original manuscript never included the whole first chapter called "The Roswell Desert." That chapter includes first-person descriptions of the event that, if authentic, could only have been written by a person present in Roswell, New Mexico, at the time of the recovery. Phil never made a claim that he was ever at Roswell during the time of the incident. Later, this rough manuscript was posted on the Internet.

Corso's most fundamentally important claim was that ET technology assisted in many of our scientific and engineering advances. This assertion is fatally flawed. According to the story we heard from Corso, some of the material from the Roswell crash was held in the Pentagon. More specifically, some of the pieces were held in his personal safe. From this office, key research and development staff members, including Corso, kept track of advanced projects ongoing in laboratories across many areas of the civilian sector. Then, once a project had demonstrated a certain level of maturity, ET material would be covertly slipped to the private company under a cover story that it was really advanced Soviet R & D that we had acquired. In other words, the scientists involved in continuing the research would not be knowledgeable of the true provenance of the items they received. The intent of this operation, Corso stated, was to enhance the American lead in various technologies while protecting knowledge of the source.

Those are all interesting ideas, but they fly in the face of the known history of technology developments, which are extremely well documented. From the beginning of our discussions with Corso we explored each of the major technological areas that he indicated had benefited from this ET infusion. When it came to integrated circuitry, Jacques Vallee was very well versed in the topic and stated Corso's assertions did not fit the historically documented facts. Similar evaluations were done for developments in lasers and fiber optics. The scientific advances in all areas ran in a linear fashion and no step functions, or unexpected leaps, were noted. We studied the established time lines provided for each technology as they are meticulously recorded and even prizes awarded in some instances. Even the most generous suggestion is that if ET technology was surreptitiously interjected, it did not do much good.

One area that I had direct access to was that of night vision. According to Corso, the concept for new image-intensifying night vision equipment came from the autopsy of alien eyes. That is simply not true. During World War II considerable effort went into the development of infrared (IR) sensors. The problem with the old IR vision equipment was that it required an emitting source. When illuminating a target, the source could easily be identified and attacked by the enemy. When photomultipliers came along, the game changed. They took existing ambient light and amplified it and by the Vietnam War we already had first-generation devices in the field. The PVS-2 we had at Ba Xaoi, my Special Forces camp in the Mekong Delta, was one of our most prized possessions. We also had a Flashy team assigned, which consisted of a searchlight with an IR filter. They were big, fixed to their location, and sent out a beam that could have been spotted miles away. For those of us who fought at night, the new ambient amplification system was a giant leap forward—but one that the genealogy of was totally accounted for.

During my time at INSCOM I had worked with what was then known as the Night Vision Laboratory on several occasions. Located at Fort Belvoir, Virginia, they had constructed some very special equipment for me and I had become friends with the director, Dr. Lou Cameron. In fact, Cameron had actually built the Night Vision Laboratory (NVL) many years before, and had intimate knowledge of the developments there. By the time we met Corso, Cameron had retired, but I was able to locate him and discuss the assertion. Lou flatly denied any involvement of alien eye coverings in the development of any night vision system, and he would have known.

In 1996 the NVL director was Dr. Rudy Buser. I had also known Rudy from my INSCOM days and arranged to go to the lab and discuss Corso's claims. Like Lou Cameron, he knew that I played in some strange areas. After listening intently to the claims, he resisted laughter, but did state that he knew nothing about anything I mentioned. He offered me a written copy of the history of night vision that was quite detailed. In these documents was another problematic issue. One of Corso's specific claims was that Trudeau had sent him to the Night Vision Laboratory with a five-million-dollar check in his pocket. The notion was that Corso, a nonscientist, was to evaluate the technical progress of the laboratory, and if he approved, Corso would hand them the check. As a former senior staff officer who had been involved in many financial transactions to boost technology, that simply did not make any sense to me. That was not the way business is conducted. Government money, when approved for transfer, is moved electronically, not by check, and certainly not "carried in my back pocket." Given that five million dollars was a substantial sum at that time, it was extremely unlikely that a nontechnical lieutenant colonel would be allowed to make such an arbitrary decision. The facts support this assertion. The published history of NVL does mention the five-million-dollar increase for their research at that time. It included the name of the person who had approved that transfer. That name was not Corso.

The Published Book Was at Variance with the Manuscript

In the original manuscript there are episodes that are even more incredulous. These do coincide with the versions of incidents that Phil verbally relayed to us on several occasions. One included an encounter with an alien while he was assigned at White Sands Missile Range (WSMR) in New Mexico. Interestingly, this site is not far from the original Roswell site. For those not familiar with the area, WSMR encompasses a huge amount of land, almost 3,200 square miles not counting Fort Bliss, which is located immediately to the south. It is the largest military installation in the United States. Generally remote, on the northern end is Trinity Site, the spot where the first atomic bomb was detonated.

The incident took place in 1957, prior to Corso's assignments in Washington. He was stationed at White Sands Missile Range, and frequently traveled across the remote areas. Corso told us that one day he had been driving alone over a desolate area in the intense desert heat. Spotting a cave—actually an abandoned mine—and seeking a place to cool down, he decided to enter it. While resting there, he said he had a face-to-face, fully conscious encounter with an alien from outer space. According to Corso, the UFOs of the time had problems navigating through our radar systems. While these UFOs were capable of traveling through the extreme radiation found in outer space, human radar systems could interrupt their propulsion systems, leading to crashes.

Corso states he and the alien had an exchange. He indicated that no words were spoken but rather information was transferred mentally. The alien then said he wanted to leave but was concerned about the radar interference. The alien asked Corso to order the radars covering White Sands Missile Range to be shut down for a brief period. Corso asked the alien what he would give him if he complied. The reported response was, "A new world if you can take it." Corso told that story frequently, and said he often contemplated what the alien meant by that statement but was never quite sure he understood. The original manuscript included drawings of the alien that Corso made after this encounter.

There were other events omitted as well from the final text. One incident included what Corso believed to be evidence for time travel. In another he reported that a chapel had disappeared from White Sands Missile Range with no rational explanation. He indicated that since the evidence to support these unusual events was so scant he made a conscious decision not to make any official reports at the time they occurred. That was probably a prudent move from a career perspective, but does leave others wondering about the veracity of those tales.

Questions Unanswered

There are conflicting stories about the editing of *The Day After Roswell*. Corso stated that he had only twenty-four hours to complete the editing of the entire book. From personal experience a single day would be an impossibly short time for any adequate review. Other people involved in the process say that was not the case and that the manuscript had been gone over page by page. However the poor editing came about really doesn't matter; the book is replete with errors as are noted in Appendix A. During our last meeting, just before his second and massive fatal heart attack, Phil and I spent most of a day together at his son's new home. We enjoyed a lengthy conversation and a leisurely lunch at a local Chinese restaurant. Both my son Mark and wife, Victoria, accompanied us for the entire period. In discussing the book, he admitted there were many mistakes that he had spotted too late in the process of publication. Phil mentioned that he was already engaged in writing another book to set the record straight. To support that statement he showed us the new handwritten manuscript that he had begun compiling.

The conundrums are obvious. There is no doubt that Lieutenant Colonel Corso was involved in various high-level positions and was assigned where he said he was. There are critics who have parsed Corso's assignments rather finely, and then extrapolated from minor nitpicks to calling him a fraud. Despite where his office might have fit organizationally at the time, my investigation of his background did put him as a staff officer very close to the White House. There is no doubt that he had some powerful friends, including Senator Strom Thurmond, who wrote the first foreword to Phil's book. Later Thurmond retracted that foreword as controversy began to rise.

The one question I could never answer was, "Why would he make up these stories?" One thing that did impress me during our many conversations was the continuity of his narratives. Usually when someone has made up a story from whole cloth, they get tripped up when retelling the story over time. With Phil, it was almost as if the incidents were recorded on tape and you could push the start button at any time and hear the identical report. Phil was very likable and seemed to be generally a credible man with an outstanding career. Unfortunately he was contradicted by the facts.

Corso had contemplated other books based on his life. One was to be *The Day After Rome*, describing in more detail his exploits with the Vatican, arranging for safe passage for 10,000 Jewish refugees to the British Mandate of Palestine, and helping establish security in post–World War II Rome while confronting the communist movement emerging there. Then there was to be *The Day After Dallas*, in which he planned to address his investigation of the JFK assassination.

Summary

Phil Corso's book is significant if only for the number of copies sold and the influence it has had on the UFO field. Even a few senior government officials not associated with UFOs came to believe his story. There is no doubt that he had an illustrious military career, which brought him into contact with many people of historical record. However, in *The Day After Roswell*, Corso made many extraordinary claims, especially regarding the use of crashed UFO material in the development of American advanced technology. Unfortunately, none of those claims have been substantiated, and most are directly refuted by known facts. It is also unfortunate that he did not live long enough to produce his response to the multitude of questions raised about the veracity of his statements.

The gulf between the public's opinion about UFOs and what most scientists believe regarding them is cavernous. The next chapter explains how that came to pass. Inauspiciously, it is a story of scientific deceit and a harbinger of the distrust that now permeates debates on momentous issues such as climate change and evolution.

3

THE CONDON REPORT REVISITED

Caution should be exercised in accepting scientific belief as scientific data.
—MICHAEL SABOM, M.D.

The single most important document ever written about UFOs was the *Scientific Study of Unidentified Flying Objects*, or, as it is better known, the Condon Report. This is stated in the same context that *Time* magazine selected Adolf Hitler as their Person of the Year in 1938, followed by Joseph Stalin in 1939 and 1942, and Ayatollah Khomeini in 1979. To be significant does not mean that their contributions were positive. In all of the examples cited, the choices were decidedly negative. The Condon Report was no exception.

The impact of this study, chaired by Dr. Edward U. Condon at the University of Colorado, has been both profound and widely misunderstood by skeptics and UFO devotees alike. The most fundamental issue that needs to be examined is, "What questions was the Condon study asked to address?" Contrary to popular belief, Condon was not asked to determine whether or not UFOs were real. Rather, there were two distinct issues involved. Like every study before this involving the U.S. Air Force, the primary question was, "Do UFOs constitute a threat to national security?" The second, and less emphatic, topic was whether or not science could be advanced significantly by studying UFOs.

Despite all of the problems associated with the Condon Report, it must be recognized that for the primary question—national security—Condon appears to have been right! In the more than forty years since the report was published, there have been no overt invasions by aliens from outer space—even though the subject has been a staple of Hollywood for the intervening decades.

For the convenience of any reader interested in the original sources, they are cited throughout this chapter, and the page numbers listed at the end of appropriate paragraphs. These pages refer to the book *Scientific Study of Unidentified Flying Objects*. Since the use of common sense throughout this book is implored, it is highly recommended that you cross-reference the material for yourself.

Historical Perspective—No Threat

Many UFO buffs erroneously assume that since UFOs by definition fly, therefore the U.S. Air Force would have responsibility for the topic. This is not true. The Air Force is under the Department of Defense, which is tasked specifically with national security. This extremely important factor has been emphasized in every study or report. Therefore, it is worth examining the verbiage used in the tasking statements of several of these UFO reports. In short, what were they asked to do? This should help the reader to better understand what was really being requested when the contract for the Condon study was established and written.

Even before the U.S. Air Force came into existence and was formally recognized as a separate branch of service under the Department of Defense and came into existence, there was interest by their aviators in UFOs and concerns related to national security. The U.S. Air Force was officially formed on September 18, 1947, and one of the first topics raising concern was that of UFOs. There is a letter dated September 23, 1947, from General Nathan Twining that was still addressed to the Commanding General of the Army Air Forces. This letter is often cited in UFO studies and it is quite significant. In that letter he gives the official opinion for Air Materiel Command that "Flying Discs"—as they then were called—were *"something real and not visionary or fictitious."* After outlining what had been observed and reported regarding the flying discs, General Twining went on to explain issues that required consideration. His first thought was *"the possibility that these objects are of domestic origin—the product of some high security project not known to AC/AS-2 or this Command."* He next addressed *"The lack of physical evidence in the shape of crash recovered exhibits"* and *"The possibility that some foreign nation has a form of propulsion possibly nuclear, which is outside our domestic knowledge"* (pages 894–95).*

What General Twining stated is very important. Also significant is what does not appear in the letter—the ET possibility. The physical reality of

---

* To assist anyone interested in checking the documents listed, page numbers have been inserted for each citation. These page numbers refer to the book by Edward Condon, which is the complete commissioned report for the U.S. Air Force, *Scientific Study of Unidentified Flying Objects* (New York: Bantam, 1969).

UFOs is assumed. However, he begins the recurring theme that some other nation may have achieved a breakthrough in aviation technologies. While not explicitly identified in the letter, this could only be their concerns about the advances in technology by the Soviet Union. This period was, after all, the initial phase of the Cold War and trust of the USSR was at its nadir. Noteworthy is that every senior official I met with who was alive during that period addressed the same assumption; if it's not ours, it must be Soviet. The prospect of a Soviet technology breakthrough would have been terrifying to those charged with our national security.

Also worth considering is that General Twining specifically addresses the lack of material evidence. Remember, this letter is dated only two months after the supposed Roswell incident and Twining's name is often associated with that event. If Roswell were a real crashed UFO, it seems more likely that he would have just not mentioned the topic at all, rather than draw attention to it. That position can be debated, but the commonsense answer should prevail.

However, the general also raises a point that comes up time and time again. That is the supposition that some other organization in the United States has a secret program and has achieved a technological leap. When considering General Twining's position as the head of Air Materiel Command, the probability that he would be kept out of the loop of a crashed UFO is vanishingly small. It is his AMC that owns all of the USAF aviation R & D capability. At that point in American history, the CIA was not in the aircraft business. Unless he suspected the U.S. Navy, with its own aviation units of involvement, the alternative organizations with advanced R & D capability are pretty few and far between. This thought points to another recurring theme: It's always somebody else in the government that is responsible (but we don't know who).

It is also important to note that the ET hypothesis is not mentioned at all in his letter. The absence of commentary concerning extraterrestrials is remarkable. If General Twining were aware of either crashed UFO material, or recovered alien bodies from the Roswell site, then a response to his superiors would certainly have included ETs as an alternative.

There Were Several UFO Studies Before Condon

Predicated on official interest, on December 30, 1947, Project Sign was established to collect, correlate, evaluate, and act on UFO-related information. The project was to examine, "all phenomena in the atmosphere which can be construed to be of concern to the national security." This directive was sent to the commanding general at Wright Field, Ohio, for implementation. (Wright Field later became known as Wright-Patterson Air Force Base.)

In February 1949 *Project Grudge* succeeded *Project Sign*. Officially it only lasted until December of that year. In the closing report it was recommended that future study of UFOs be reduced in scope. The conclusions noted, "There is no evidence that objects reported upon are the result of an advanced scientific foreign development; and, therefore they constitute no direct threat to national security" (page 509).

Project Blue Book, the U.S. Air Force official program to study UFOs, was reviewed in 1966. *The Special Report of the USAF Scientific Advisory Board Ad Hoc Committee to Review Project "Blue Book"* concluded that "in nineteen years since the first UFO was sighted there has been no evidence that unidentified flying objects are a *threat to our national security*" (emphasis added; page 813).

In the tasking memorandum for Military Director, Scientific Advisory Board, the Office of the Secretary of the Air Force noted that the study was, "In keeping with its air defense role, the Air Force has the responsibility for investigation of unidentified flying objects reported over the United States." Please note it does not address UFOs in other parts of the world. While supportive of continued investigation, the memo stated, "To date, the Air Force has found that *no evidence of the UFO reports reflects a threat to our national security*" (emphasis added). The same memo did note that many sightings from veridical sources could not be explained (page 816).

Air Force Regulation No. 80-17 dated September 19, 1966, entitled, *"Research and Development, Unidentified Flying Objects,"* established the program to investigate reports and indicated compliance was mandatory. This again specifically was based on their role in air defense. The program objectives stated, "Air Force interest in UFOs is twofold: to determine *if the UFO is a possible threat to the United States* and to use the scientific data gained from the study of UFO reports" (emphasis added). The regulation goes on to note that most UFOs have a conventional explanation and *"present no threat to our security."* It also addresses the possibility *"that a foreign nation may develop vehicles of revolutionary configuration of propulsion"* (page 820). The regulation does not include the possibility of extraterrestrial craft being involved.

The UFO topic continued to garner popular interest and in January 1953 the CIA convened a secret panel to review the topic. The Scientific Advisory Panel on Unidentified Flying Objects, better known as the *Robertson Panel*, was comprised of several prominent scientists including Howard Robertson, Luis Alvarez, Thornton Page, and others. There had been some significant sightings, including a dramatic case over the nation's capital and there was concern about how the public would respond to future cases. Specifically they were worried about mass hysteria. For a few days the panel reviewed cases and had briefings from the head of Project Blue Book. Also present was the later famed Dr. J. Allen Hynek.

Unfortunately the panel went far beyond benign observations in their report. Of course they addressed the threat issue and concluded, "That the evidence presented on Unidentified Flying Objects shows *no indication that these phenomena constitute a direct physical threat to national security*" (emphasis added). They went on to state, "We firmly believe that there is no residuum of cases which indicate phenomena which are attributable to foreign artifacts capable of hostile acts, and there is no evidence that the phenomena indicate a need for revision of current scientific concepts" (page 918).

However, the panel went on to suggest debunking UFO events so that the public would lose interest in the topic. Concerned about a strong psychological reaction, they advocated an educational program that would be carried out by the mass media and aimed at reducing the gullibility of the public. The panel members also expressed concern that an enemy (unnamed) might possibly use UFOs as a form of psychological warfare and that the reporting systems could be overloaded (page 915).

The Colorado Study—Damage Done

By 1968 the Air Force, as an institution, was quite tired of UFOs, and really wanted nothing to do with them. Despite the fact that periodic observations of remarkable events were made by some of their own personnel, the senior leadership just wanted to have the topic go away. There is a management adage that states an action passed is an action handled. This means, if you can get someone else to take responsibility, then you have accomplished your task. It is also well known that the one way to kill such a task is to study it to death. That usually means appearing to look busy but without doing anything concrete about the real problem. Thus the Air Force established the Condon study with the real objective of eliminating a residual pain in their ass.

It was necessary to have an external scientific body conduct the evaluation. That provided the illusion of independence and credibility. Internally the Air Force frequently had been following the advice of the Robertson Panel, and debunking cases whenever possible. The rationale for the Air Force position was quite simple—resources. Budgets are invariably a zero-sum game. If you have personnel working UFOs, then they are not doing something else. In this case they were not contributing to the Cold War effort, or supporting the war in Southeast Asia, which by this time was getting into full swing.

Condon was just what the doctor ordered—he had the outstanding scientific credentials and a mind-set that supported the outcomes desired by the Air Force. While many UFO supporters today think the study was easily assigned, the reality was that several institutions turned down the opportunity. The money, eventually increased to a little over $500,000, was deemed not worth the potential risk of damage to their institutional credibility. Fortunately for the Air Force, the University of Colorado was willing to step up to the plate.

The Condon Report Was Not All Bad

Ironically, parts of the Condon Report are pretty good. As with Project Blue Book, the investigators described cases that defied conventional explanations. While the majority of all UFO cases had prosaic answers, often misidentification of known artifacts, there always remained a core set of events with high credibility and high strangeness. The problem with the report is internal incongruence between the *Conclusions* and *Recommendations*, which were written by Condon himself, and the body of the study. While Condon never personally investigated any of the cases, he simply chose to ignore the analytical body of his own report written by his experts. The conclusions, like every study before this one, stated they did not find any indications of a defense hazard. He went on to affirm the prior positions taken in earlier studies stating, "We know of no reason to question the finding of the Air Force that the whole class of UFO reports so far does not pose a defense problem." Obligingly he went on to recommend the termination of Project Blue Book, saying there was no need for a special organization to study UFOs.

But there was more damage to come. Condon took on the issue of possible scientific advancement and basically laid it asunder. He wrote, "Careful consideration of the record as it is available to us leads us to conclude that further extensive studies of UFOs probably cannot be justified in the expectation that science will be advanced thereby." Addressing astronomers, atmospheric physicists, chemists, and psychologists, he said they have "decided that UFO *phenomena do not offer a fruitful field in which to look for major scientific discoveries*" (emphasis added; pages 1–2). Further he noted they should turn their valuable attention and talents elsewhere.

Following the lead of the Robertson Panel, Condon also addressed education. He concluded that, "We feel that children are educationally harmed by absorbing unsound and erroneous material as if it were scientifically well founded. Such study is harmful not merely because of the erroneous nature of the material itself, but because such study retards the development of a critical faculty with regard to scientific evidence." He went so far as to recommend that teachers should refrain from giving students credit for schoolwork based on UFO material (page 5).

There Was Dissention on the Panel

It is well known that there was dissention within the committee. Two of the members, Dr. David Saunders and Dr. Norman Levine, were fired for "incompetence." However, the real reason had to do with a controversial memo written by Condon's chief assistant, Robert Low, that thoroughly undercut the efficacy and scientific standing of the study.

Sometimes referred to as the "Trick Memo," in part it stated: *"The trick would be, I think, to describe the project so that, to the public, it would appear a totally objective study but, to the scientific community, would present the image of a group of nonbelievers trying their best to be objective but having an almost zero expectation of finding a saucer. . . . One way to do this would be to stress investigation, not of the physical phenomena, but rather of the people who do the observing—the psychology and sociology of persons and groups who report seeing UFO's. . . . If the emphasis were put here, rather than on examination of the old question of the physical reality of the saucer, I think the scientific community would quickly get the message."*

Basically he told people to look busy, but since the outcome was already predetermined, they were really focused on deceiving the American people, thus providing the U.S. Air Force what they really wanted—out.

David Saunders went on to write his own book, *UFOs? Yes!* As a psychologist he had been brought into the study to prove that the people who saw UFOs were less than mentally stable and so their observations were unreliable. Previously he had worked extensively with the CIA and was well established in the field of psychology. Among his attributes was an understanding of group psychology and he had developed an instrument that measured how well groups interacted. He could also determine why groups were dysfunctional based on the internal variability of the participants. After testing a substantial number of people who had reported observing UFOs, he concluded that as a group they were more sane and balanced than any other group he had ever tested. This finding undermined one of the presuppositions of the study, which was that many of the people who reported UFOs were not mentally stable. Unfortunately the negative concepts that observers had psychological or drinking problems slipped into public consciousness. This had a dampening effect on high-quality UFO reporting. To address his findings, we had the good fortune of having David Saunders appear before one of the key science groups arranged for the ATP study.

Report Analysis

In evaluating the Condon Report, it was striking how the format of the study differed from most others. Fundamental to most scientific studies are the terms of reference or TOR. These are critical as they establish the boundaries of the study and are usually stated succinctly at the beginning of the report. Without well-constructed TORs, studies usually flounder, as happened with this study. Also missing is a traditional Executive Summary, which is designed to convey the essence of the report to extremely busy senior leaders. Essential to most study reports, these are one or two pages in length, even for the most complex of topics. Longer than that, and they won't get read by decision makers. Instead, Condon opened with his personal conclusions and recommendations, using a "royal we" to convey the illusion of broad support for his statements. This is followed by a rambling forty-plus page summary, which would rarely be read by executives. This summary is either misleading, or points to serious intellectual limitations in the participants. He concludes by stating, "In our study we gave consideration to every possibility that we could think of for getting objective scientific data about the kind of thing that is the subject of UFO reports." If this is true, why then did Condon disregard some of the most spectacular cases, ones that included multisensory data and high-quality witnesses? The best example is the evidence of the RB-47 electronic warfare aircraft case that lasted for about ninety minutes over three states and points to capabilities that cannot be duplicated today. There will be more on that case later as it leads to another, previously publicly unreported observation.

In the body of the report there are a number of cases that certainly support the notion that UFOs were real and interacting with national defense systems. At least two are listed as real UFOs. Here are a few examples, with the committee conclusions:

> — A crew of a C-47 saw a UFO from above and below. They maneuvered for ten miles but could not shake it. The sighting was confirmed by ground radar. Observations include acceleration from 0 to 4000 MPH instantaneously and a right-angle turn at 600 MPH. Conclusion: Real UFO.
>
> — Six witnesses, including two police officers, observed an aluminum-looking metal disc the size of an auto at a height of 50 feet above a school. Observation lasted

30 to 45 minutes. The UFO changed color and sped off.
Conclusion: no explanation.
— A USAF pilot with "irreproachable background" took
photos of a UFO as he flew toward the Rocky Mountains. Two photos showed domed object. Conclusion:
Claim not supported.
— A nuclear physicist and two others saw a bright object below
the clouds at 300 feet. The colors changed from red to
brilliant orange and the UFO was so bright it washed out
his headlights. He estimated power at 500 to 800 mega-
watts. Conclusion: Unknown but doubt the intensity.
— A farmer and one other photographed metal object about 20
to 30 feet in diameter. Object went from hover to high
speed with rapid change in direction. The photos were
not fakes. Conclusion: Extraordinary and all evidence
fits.
— Three people observed a small shiny object dropped from
larger UFO. It had trouble maneuvering and molten
metal was seen dripping from UFO. Conclusion:
Canadian Air Force says real. Impressive testimony.
— Two USAF guards report UFO near a missile base. Sighting
confirmed by radar. Fighters scrambled but did not
contact UFO. Conclusion: Not sufficient evidence.

Despite cases like these, Condon wrote conclusions that in effect said there
was not scientific value in understanding rapid acceleration and high g-force
maneuvering. The case listed last had a most interesting caveat. This report
was made during what is known as the Northern Tier sightings. These were a
large number of sightings across the northern American missile defense systems. The note stated the reason this report was made was that the guards
were new and "most guards don't report UFOs." The implication was they
were so common as to not attract the attention of the security patrols. Still,
Condon did not think this was a defense matter.

There Were Immediate Rebuttals

While many of the UFO researchers of the time initially supported the Condon study, they quickly became disillusioned and realized where the report
was headed. One of those was Dr. James McDonald, a highly qualified atmospheric physicist working at the University of Arizona, Tucson. McDonald

had developed an interest in the topic and been involved in credible research concerning UFOs. There are erroneous reports published that McDonald was a member of the Condon Study. In fact, he attempted to gain a position on the study, but was denied.

McDonald was one of the first scientists to address the major flaws in the study. His analysis was reported in UFO circles as he was extremely critical of the incomplete scientific analysis of many UFO cases cited by Condon. He noted that the Condon study intentionally addressed cases that were relatively trivial, and often did not include the complete details. His article, "Science in Default: Twenty-one Years of Inadequate UFO Investigations," appears as a chapter in *UFO's: A Scientific Debate*, edited by Carl Sagan and Thornton Page. Addressing Condon, and prior efforts to debunk sightings, McDonald wrote, "Prior to my own investigations, I would never have imagined the widespread reluctance to report an unusual and seemingly inexplicable event; yet that reluctance, and the reluctance of scientists to pay serious attention to the phenomena in general, are quite general." McDonald then went on to cite specific cases inadequately addressed in the Condon Report.

Another senior scientist with exemplary credentials, Dr. Peter Sturrock, an astrophysicist at Stanford University, attempted to publish a report rebuking the Condon Report.* Sturrock obtained his doctorate from Cambridge University and by the time of the Condon study had already made a number of inventions, one of them being a microwave tube that operates on a principle subsequently named the "free-electron laser." The point is that Sturrock was well established and not a scientist that could easily be ignored. Yet, that is exactly what happened. The details are described in his most recent book, *A Tale of Two Sciences: Memoirs of a Dissident Scientist*. In a chapter devoted to the Condon Report, Sturrock covers the in-depth analysis he conducted of Condon's study, and his efforts to get them published.

Like McDonald, Sturrock noted the inadequate selection of cases and incomplete details provided. He indicated that although the case studies should have represented the heart of the report, they accounted for less than one quarter of the pages. He also noted that about half of the 900-plus page report had little or nothing to do with the topic of UFOs. Despite extensive detailed analysis and appropriate credentials, Sturrock was turned down by every mainstream scientific journal to which he applied. By establishing an organization devoted to

---

* For full disclosure it should be known that I have been acquainted with Peter Sturrock for more than two decades and currently serve as a councilor of the Society for Scientific Exploration that he helped found. I consider him to be a friend.

studying anomalous phenomena, Sturrock did get his message across. His report, "An Analysis of the Condon Report on the Colorado UFO Project," is available on the Web site of the Society for Scientific Exploration.* As late as 2002, Sturrock attempted to enter a debate in *Physics Today* that contained laudatory remarks about "Condon's thorough analysis of the UFO Problem."

Therein lays a fundamental problem with the Condon Report that continues to this day. It was the conveyance through both mainstream media and scientific journals that the Condon study had completed an extensive scientific analysis of the data. They had not. Condon came to his conclusions before the study began, and proved to be disinterested in facts, even those contained in his own document. While there were prompt responses decrying the poor quality and incongruence of the report, they were drowned out by the more prominent establishment publications such as *Nature*.

As determined from various scientific briefings that I gave on UFOs, nearly all of the scientists who had not personally investigated the topic assumed that the Condon study was the final word. The most frequent response from those scientists was, "That's already been done!" It is not uncommon for skeptics to point to the study and proclaim there is no evidence. They are really stating they have no idea what evidence might be available, and are not about to review any data. After all, Condon gave them a pass.

Along Comes Carl Sagan

Not surprisingly, Carl Sagan and Thornton Page, in their editors' introduction to the book *UFO's: A Scientific Debate*, reinforced this notion of thoroughness. The book followed shortly after the Condon Report. In it they called the report "one of the most detailed examinations of the subject ever performed." They lamented that "young people are finding science increasingly less attractive" and stated, "We all agree that this drift is deplorable." They go on to express concern that these same people have interest in "borderline subjects" including "UFOs, astrology, and the writings of Velikovsky." The dignity and interest afforded UFOs had been brought about, they claimed, through "widespread newspaper and magazine coverage which reaches many more Americans."

These comments sound very similar to the views expressed by Condon regarding restricting the UFO topic from education. They accurately portray the sentiments of Carl Sagan and are far different from the congenial public demeanor he wished to convey to the public. That fact was vividly revealed in a

---

* www.scientificexploration.org/journal/jse_01_1_sturrock_2.pdf

meeting at Cornell University in 1988. Since the annual meeting of the Society for Scientific Exploration (SSE) was held at the university where he worked, the courtesy of an invitation to speak was extended to Sagan. He accepted and gave one of the most amazing talks I have ever heard. In keeping with his veiled tendency toward scientific elitism, Sagan informed the audience that technical debate among highly educated people with proper training in critical thinking should be permissible. (For perspective on the SSE, consider that at that time full membership required a doctoral degree and experience in your respective field.) While we were properly anointed and could conduct critical analysis regarding anomalies, there was danger, he suggested, with public discourse on these topics. Discussions about phenomena including UFOs were, Sagan contended, undermining not only public understanding of science, but the foundations of democracy itself. If the great unwashed were exposed to these topics, educational chaos could follow. His comments, synthesized down to the most basic level would be, "Freedom of speech will be the downfall of democracy."

Scientists and the Public Are At Odds — a Significant Problem for Science

This recurring theme that UFOs, and other reports of phenomena, constitute a threat to education is worthy of public concern. As has been shown in various studies, there is a vast chasm between the belief systems of those who identify themselves as scientists and the general public. The problem facing science is that a very substantial percentage of the population has had personal experiences, often at a significant emotional level, that run counter to the beliefs of science in general. The basic premise of many scientists regarding phenomena is: it can't be true, therefore it isn't. As previously indicated, surveys continue to show that about 7 to 10 percent of adults have experienced the sighting of what they believed was a UFO. Undoubtedly, many of those observations are misinterpretations of the event. However, when you have multisensory data, multiple credible witnesses, and close encounters that last for significant periods of time, or recur frequently, they should not be categorically denied—as is the norm.

A huge issue for science and scientists in gaining public acceptance is that they have been proven wrong on numerous occasions. One has but to have followed the conflicting changes to seemingly dogmatic strictures on what is dietetically good for you to understand the problem. In the introduction to his UFO book Sagan noted that, "Science has itself become a religion." Its adherents certainly behave as if that is true. UFOs, and other phenomena, if proven to be real, constitute a serious threat to their psychological makeup as they contradict the very foundation of many scientists' belief systems. They seem to ignore the fact that big changes in science usually come, not from incremental expansion of knowledge, but exploration of anomalies that seem to defy theoretical predictions or by accident. There is a litany of such inventions. Penicillin probably tops the list, but other unanticipated examples include vulcanized rubber, radioactivity, plastic, and Teflon. It may also surprise readers to know that Viagra was discovered accidentally as an unanticipated side effect when the researchers were attempting to make a new angina drug.

That same introduction indicated a "*belief that the public understanding of science was at stake and the borders between scientific and nonscientific discussion need explicit delineation.*" If they were worried that UFOs would upset the scientific applecart, they did not envision what science would do to issues like global warming or intelligent design.

It might astonish some people who have read Sagan's book, *UFO's: A Scientific Debate*, to know that it is actually quite good in parts. If nothing else it articulates the key differences in arguments between UFO proponents and opponents. In the book, the proponents discussed facts, while the opponents relied on emotion to stress their respective positions. Researchers such as Hynek and McDonald addressed specific cases and provided details. Opponents, such as Lester Grinspoon and Alan Persky, indicated that both UFO observers and interpreters exhibited unusual emotions and indicated the topic raised a fervor normally reserved for religion or politics. This is in obvious contradiction to what Saunders had found when conducting personal interviews with these same people.

Most of the readers will be familiar with Carl Sagan's famous statement about billions and billions of stars and indicating that based purely on probabilities, intelligent life somewhere else in the universe cannot be ruled out. He also went on to state that there is not a bit of scientific evidence to support that we may have been visited. After listening to him speak and reading the book he reportedly edited, it seemed clear that Carl Sagan did not even read his own material. Certainly, some hard evidence was reported, even if it was not conclusive.

The Outcome

The negative impact of the Condon Report on the field of science cannot be overstated. Although it was clearly an inferior research effort, the wide support for the conclusions conveyed via both scientific and popular journals has dampened enthusiasm for further serious research projects. Anyone associated with UFO research is well aware of the paucity of funding and the potential for risk to their professional standing if they openly venture into the field. This continues to this day, and most of my senior scientist friends who are interested in obtaining quality data on the topic do so with an understanding of anonymity—and with good reason.

The risks include guilt by association. My personal examples include ad hominem attacks by both *Scientific American* and *The Bulletin of Atomic Scientists*. John Horgan, addressing me and writing in the July 1994 issue of *Scientific American* stated, *"His 'interest' in alien abductions and paranormal phenomena, about which most scientists are deeply skeptical, raise questions about his judgment and is therefore a legitimate part of the story."* In other words, Horgan was stating that if scientists did not go along with the herd, they don't think clearly.

Similarly, in the September-October 1994 issue of *The Bulletin of the Atomic Scientists*, Steven Aftergood wrote, *"John Alexander is by all accounts a resourceful and imaginative individual. He would make a splendid character in a science fiction novel. But he probably shouldn't be spending taxpayer money without adult supervision."* What is most significant about these statements is that the topic of the articles had nothing to do with phenomenology. Rather, the stories covered research efforts on nonlethal weapons as life-conserving alternatives on future battlefields. These authors were opposed to this effort and chose to attack me personally rather than address the real issues.

The bottom line for both observers of unusual events and UFO researchers is to acknowledge their vulnerability. The Condon legacy lives on and it is little wonder that people are reluctant to come forward—especially if they have something to lose.

Summary

The Condon Report is important to the study of UFOs for the lasting damage that it has done. Despite considerable evidence that the study was not scientifically sound, it is the perception that it was that has lived on for more than four decades. It is also important to remember that the main question to Condon was whether or not UFOs constituted a threat. His assertion that they were not seems to have been substantiated. However, his commentary about what science might learn from the study of the phenomena and claims that even considering the topic was detrimental to American education were far off the mark. The bottom line was that the report accomplished its mission; it allowed the U.S. Air Force to drop their requirements to investigate UFOs.

4

THE CONGRESSIONAL HEARINGS THAT NEVER HAPPENED

Courage is knowing what not to fear.

—PLATO

The ultimate goal for many UFO disclosure enthusiasts is to have Congress hold hearings. There it is assumed the truth will suddenly pour forth and nefarious elements will emerge from the shadows and confess that they have been withholding information that ET is among us. Unfortunately, this is a very naïve position, as most investigative Congressional hearings go nowhere. They are useful for getting material established on the record, but rarely lead to action. One has but to read the *Congressional Record* to find the amount of material that Congress generates. Still, having hearings would be better than nothing at all. As a minimum they could have provided a touchstone that could be referenced and used to support further endeavors aimed at getting UFOs as an acceptable topic for mainstream science.

Given the cacophony for Congress to hold hearings regarding UFOs, it is important to understand their function, and how they might come about. The reality is their impact would be decidedly more limited than is generally realized. It is also extremely important to comprehend the über-sensitive political environment that can stigmatize any public figure who even suggests they support the notion of UFOs.

In the case of such hearings, the adage *"Be careful what you wish for"* is applicable. Done properly, the outcome could establish a positive benchmark. However, given the resistance and resentment of many scientists to the topic of UFOs, there is a strong probability that there would be negative consequences. Those findings would come from die-hard skeptics who would be invited on the premise that the committee wanted a balanced report. The skeptics would have a tremendous credibility advantage as they regurgitate Condon and simply state that there is no hard evidence. They never let facts get in the way of an opinion. If the skeptical position prevailed the results could be devastating.

Hearings on UFOs have been once-in-a-generation events at best. The yield cannot be left to chance. To obtain positive outcomes from Congressional hearings on UFOs would require a lot of very carefully laid groundwork, which would have to be quite different from just UFO enthusiasts flooding their representatives with e-mails, faxes, and calls requesting that hearings be held. In a politically sensitive environment one should never ask a question to which you don't already know the answer. To accomplish that would require that many people be briefed before the formal session. They would include both staffers and principals. It would be essential to know if there was anyone who would use the sessions for political grandstanding in order to hurt members of the opposing party. Only after consensus is assured should a public meeting be held.

The reality is that there was a plan to hold such hearings. Why they fell apart will be explained shortly, but the blame rests solely with UFO protagonists. In the summer of 1999 I was approached by the CEO of a small defense contractor who mentioned that a Congressman wanted to have fact-finding hearings about UFOs. He put me in touch with a Congressional staff member who again asked about the potential for holding hearings regarding UFOs. She said the Congressman for whom she worked was interested in the topic and wanted to sponsor a one-day session. At the time he was in a position to call for hearings, but still would have to gain approval from another congressman who chaired the House Committee on Science and Technology.

Political Risks

Working part-time for NIDS, and needing some resources to make meetings happen, I contacted my boss, Bob Bigelow, and advised him what was being proposed. We then arranged for a face-to-face meeting with the congressman. That meeting was held on August 23, 1999. After introductions, the material from the Advanced Theoretical Physics briefings was covered. However, I did preface my statements with an extremely blunt question, "Can you take a twenty-point hit in the polls?"

The question was not as abstract as it might sound. Contrary to popular thinking among the UFO believers, the topic is not a voting issue for the general public. While the majority of people state they believe in UFOs, verbal support of such topics by an elected official comes with very strong negative political consequences. If you doubt that, consider the derision that presidential candidate Dennis Kucinich received after UFOs were broached during a national debate.

Dennis Kucinich Has Experienced a Tar Baby

But that was hardly the end of it. The UFO topic was brought up in a very negative light and used against him again regarding his vote on the health care reform bill debate in March 2010. When he announced that he would support the proposed bill, many critics immediately pointed out that Kucinich stated he had once seen a UFO. For example, one report noted, "Tim Egan at *The New York Times* has a terrific piece on liberal UFO-seeing Dennis Kucinich's decision to vote for health reform." Fox News doubled up on him in a March 17, 2010, article with popular host Glenn Beck quoting Tim Russert who asked Kucinich, "Did you see a UFO?" That same day Rush Limbaugh made a harsher comment equating UFOs with mental illness. Limbaugh's statement was, "There's an easy explanation for this Kucinich business. They either threatened him with a UFO trip, or they offered to make sure that in his case his preexisting condition of mental illness will not be a barrier to any kind of coverage that he gets down the road." Ken Blackwell, writing in WORLDmag .com, a publication that declares itself to profess a "Christian view," published the following comment: "Little Dennis Kucinich even got a plane ride on Air Force One. What fun. That must have been almost as exciting for Congressman Kucinich as sighting a UFO with Shirley MacLaine. He swooned and promptly announced he would vote for Obamacare." Mike Littwin of the *Denver Post* wrote, "Rep. Dennis Kucinich, who looked again at the bill and decided it wasn't a UFO." The business sector also jumped on the bandwagon. Writing in The STREET.com Matthew Buckley, a former U.S. Navy F-18 pilot and Top Gun graduate with forty-four combat missions over Iraq, had this to say about Kucinich: "This is the same representative that has personally seen a UFO, along with another member of the elite brain trust, Shirley MacLaine [*sic*]."

The reason for including so many quotes about this material is to point out just how extremely toxic the subject is from a political position. This is the proverbial tar baby that once touched will never go away. Contrary to the fairy-tale portrayal held by UFO aficionados, rather than being seen as brave defenders of truth and justice, politicians who dare to venture into this domain have entered Oz. It should be remembered that Kucinich's position was only that he saw something that was flying and he could not identify it. However logical that statement was, it is lost in the emotional clutter that extrapolates any mundane claim of a UFO sighting to little green men who are here to devour children or subjugate the planet.

Please note that this problem is not isolated to Dennis Kucinich nor are the effects restricted to one political party. Even the highly regarded and very popular President Ronald Reagan bore the brunt of considerable ad hominem attacks when it was learned that he and the First Lady, Nancy Reagan, were in contact with astrologer Joan Quigley. In his memoirs, the former White House Chief of Staff, Don Regan, exposed the secret. Regan, in the true exposé fashion of a man forced from power, declared: "I have revealed in this book what was probably the most closely guarded domestic secret of the Reagan White House." *Time* magazine quoted him as stating that he "was in a position to see how the First Lady's faith in the astrologer's pronouncements wreaked havoc with her husband's schedule. At times, he writes, the most powerful man on earth was a virtual prisoner in the White House." Clearly Don Regan knew the negative connotations that attend a politician's association with any paranormal phenomena. As word leaked out about astrology President Reagan was forced to defend his position and told *The Washington Post* that his decisions had not been influenced by the stars. Externally some were concerned and acted on their convictions. One reporter published the following: "I [Jeremy Stone] wrote to Reagan, conveying a letter endorsed by five Nobel Prize–winners, saying that as scientists we were 'gravely disturbed' that he might believe in astrology and fortune-telling. We asked that he clarify his position since we did not believe a person whose decisions were based even in part on such 'evident fantasies' could be trusted with the grave responsibilities of the American presidency." That is pretty threatening commentary, and less powerful people might have been brought down by such attacks. As will be discussed in more detail later, Ronald Reagan also had UFO sightings.

Senator Claiborne Pell of Rhode Island, who served five terms, was a friend of mine. It was well known that he had interests in various phenomena and the public response, especially from the media, was not kind. Nasty articles were written when it became known that he merely had stopped by a MUFON annual meeting that was being held in the Washington area. These despicable attacks even continued in his obituary in *The Washington Post*.

An opponent of Senator Pell's had used his inquisitive interests against him in an election campaign. The concomitant drop in the polls was about twenty points, which is where my number came from. Interestingly, it was also known that his opponent was personally favorably disposed toward these topics, but chose to use them as a calculated attack. She correctly guessed that the public would see support for UFOs, remote viewing, and the like as an indication of possible irresponsibility.

Defining Key Issues

The Congressman was very solidly supported in his district, and assured us he could withstand the pressure of the political minefield he was choosing to enter. His position established, we proceeded to provide what was believed to be the essential elements of information about what the issues should be explored. These were:

Key Issues Related to Congressional Hearings on UFOs

— There is undeniable evidence that UFOs constitute a topic worthy of serious investigation.
— Politically, the topic is a proverbial minefield.
— There is a cavernous gulf between the general population and many scientific organizations relative to their beliefs about UFOs.
— Most people believe UFOs exist and many (about 7 percent of adults) report having seen them.
— Most UFO sightings are misidentifications of known objects. However, a significant number are truly unknown objects, and sometimes they are observed at close range.
— There is strong multisensory, physical, and credible witness(es) evidence to support the existence of UFOs.
— The topic can and will become very emotional.
— Any attempt to study this topic with public funds will be actively resisted by many scientists who will paternalistically claim to speak for everyone.
— There is a very real potential for ad hominem attacks against anyone viewed as supporting this study.
— The history of the U.S. Government in regards to UFOs has not been good. They (USAF and the CIA) have consistently lied and been caught at it.
— There are a substantial number of people who are very frustrated by the government's apparent lack of interest and openness toward this topic.
— There will be a need to establish the Terms of Reference in order to bound the hearings.

The first question to be resolved should be: What are we trying to accomplish? This is not as simple as it may seem. There will be at least three distinctly different issues discussed:

1. Is there sufficient hard evidence to support a serious investigation of UFOs?
2. What does the government know about UFOs and is not telling the public?
3. Does this study include "alien abductions," "cattle mutilations," and/or other related issues of anomalous phenomena?

— No matter how hard you try, there will be people, both pro and con, who will complain about the conduct of the hearings.
— High-quality scientists and witnesses can be made available for hearings.
— There are many "nuts" and otherwise incompetent people who will attempt to get involved. If allowed to testify they will destroy the credibility of those scientists and witnesses providing hard evidence.
— The worst thing that can happen is a "whitewash." That would bring both the wrath of UFO supporters and exacerbate the opponent's attacks for conducting the hearings to begin with.
— The timing of the hearings will be critical. Careful preparation should be conducted so that they are neither a waste of time nor embarrassing. There is no need to hurry.
— The task for Congress is to appear to be responding to expressed concerns of their constituents; be neutral to the topic, and be sufficiently thorough, so as to protect the credibility of all involved in the hearings.

We should predetermine probable outcomes in descending order of preference:

1. A formal independent scientific panel be established (outside NSF) to study the issue and produce a report for Congress. (Create an ongoing active investigation.)
2. Direct that government agencies with UFO information, or that acquire such information, provide it to Congress or their appointed representatives. (Affix responsibility inside the government.)

3. A Congressional recommendation that UFOs be examined by scientific organizations. (Give formal permission to study the topic.)
4. A Congressional report be produced reporting the contents of the hearings. (Report to the public of what happened.)
5. A Congressional recommendation that nothing further be done at this time.

During the discussion that followed several administrative tasks were derived. Hearings don't just happen. There is considerable groundwork needed. Importantly, the hearings must have a charter that grants authority and delineates the purpose of the meeting. There is also a requirement for a briefing paper that provides the background information about the topic. These are usually aimed at the staff personnel who really follow the action and prepare their principals for the meeting. It was agreed that I would write the initial draft of these documents. The draft documents, including the charter, background material, and chairman's opening statement are included at the following appendices:

— Appendix B: Draft charter
— Appendix C: Background material
— Appendix D: Chairman's opening statement

It cannot be overstated that the documents are drafts that I prepared and forwarded to the key people involved. They do not constitute official U.S. Government documents. Had the hearings gone forward, each of the documents would probably have undergone several edits. However, they are presented to demonstrate that this was a serious effort that had made significant, albeit slow progress.

In our discussion we focused on specific areas that hopefully would get the attention of other Members of Congress. It was felt that issues related to interactions between UFOs and our national security were natural topics. The second issue was aviation safety. Both of these topics had tangible implications that were in the purview of Congress, and there was solid evidence to support each of them. We wanted to steer clear of generic issues such as do UFOs exist, or worse, confronting agencies that many people believe have been hiding evidence from the public. As an initial step, it would be necessary to prove there was a problem and later the topic could be expanded.

The advice to the Congressman was as follows:

— Privately make the decision to hold hearings
— Do not be in a hurry to set dates

— Increase your personal knowledge in the field

— Broaden your support base on this topic within Congress

— Allow for development of supporting materials

— Assist in dissemination of information to other Members of
      Congress and key staffers on a close hold basis

— Delay formal announcement of any decision

The Chairman Blocked Progress

There were several meetings held, including one in the Capitol in Washington, D.C. Planning progressed slowly and selected staff members were questioned about support. The main stumbling block was Wisconsin Congressman James Sensenbrenner, who held the chairmanship of the House Committee on Science, Space, and Technology. It was his committee that would have to sanction the hearings, and Sensenbrenner was adamantly opposed to the concept.

We suspected he knew nothing about the topic and his declination was a visceral response to a proposal with possible adverse consequences. Remember, with elections every two years, Members of Congress are constantly running for office. Clearly, he did not want any such hearings that might be attributed to him on his watch. An offer was made to brief him privately. Bigelow agreed to support a trip by me and Peter Sturrock to meet Sensenbrenner and provide him more credible information. While I had the information about both the prior ATP activities and the rationale for the hearings, Sturrock was an eminent astrophysicist from Stanford University. His credentials were impeccable and he had been actively involved in the study of UFOs. Among his positions he had been the chairman of both the Plasma Physics Division, and the Solar Physics Division of the American Astronomical Society. He was a founder of the Society for Scientific Exploration, and had recently conducted a meeting attended by top scientists to explore the evidence supporting UFOs. That meeting, which had been sponsored by Laurance Rockefeller, resulted in a book, *The UFO Enigma: A New Review of the Physical Evidence*, which covers a number of highly credible cases. Despite our offer, which included anonymity and confidentiality, Sensenbrenner flatly refused even to meet with us. He did not even deflect us to a staff member, which is an accepted protocol to avoid personal contact involving thorny subjects.

However, at the time we knew that Sensenbrenner was looking for a more prestigious and influential committee to chair. He had made it clear to other Members of Congress that his eye was on the powerful House Judiciary Committee. Our strategy was simple: wait until after the next election and address the hearings with the next chair of the House Committee on Science, Space, and Technology.

While we were watching and waiting for the right opportunity, on May 9, 2001, Steven Greer and his *Disclosure Project* came to town. His show produced much fanfare, ranting and raving about the terrible government agencies that were lying to the public. However the result was that his proceedings held at the National Press Club destroyed any possibility of hearings on UFOs for the foreseeable future. Those Members of Congress who quietly had been supporting the process made it known they were out. They wanted nothing to do with a high-profile circus that could have adverse impact on their potential for reelection.

The biggest problem with the Disclosure Project's hoopla was a failure to vet their performers. While there were a few legitimate observers, Greer also profiled people such as Brigadier General Stephen Lovekin, who has no U.S. military credentials at all. Since the Disclosure Project listed an Army general as a key witness, it was decided to corroborate the bona fides of Lovekin. There are rosters of all Army general officers that are distributed. The lists are complete with all active duty and retired general officers in the U.S. Army, the U.S. Army Reserve, and the Army National Guard. Lovekin was not on any list. Calls to the U.S. Army GOMO, the General Officer Management Office, confirmed the same information, as did the North Carolina National Guard where it was claimed he was assigned. With such bogus representation, presenters like Bob Salas, who really did have an extremely important experience to report, were simply overshadowed.

There Were Other Muted Attempts at Hearings

In July 2005 there was another attempt to have brief hearings that initially were to include unidentified aerial phenomena (UAP) and UFOs. According to reliable sources, the hearings were sponsored by then Congressman Tom Davis (R) from Northern Virginia. While not disclosed publicly, Davis allegedly had a sighting of his own and that is what prompted his interest in these hearings. In the preparatory discussions the focus was to be on UAP as it impacted airline safety. However, the topic merged into airline security as it was related to counterterrorism. There was a preliminary meeting held on May 23, 2003, and four UFO proponents did talk with Davis and other congressional staff members. They were Edgar Mitchell, Peter Sturrock, Jacques Vallee, and Richard Haines. The hearings were finally held on July 21, 2005, but had been taken in a totally different direction—counterterrorism. To counter any potential threat discussion, representatives from the Department of Homeland Security, DoD, and the FAA all testified that the skies were safe, but made no mention of the UAP or UFO concerns that had promulgated the hearings.

On July 25 the General Accounting Office (GAO) put out a report on the hearings now entitled, *Homeland Security: Agency Resources Address Violations of Restricted Airspace, but Management Improvements Are Needed*. The report is available on the GAO Web site. They noted that between September 12, 2001, and December 31, 2004, 3,400 violations of restricted airspace occurred; most committed by general aviation pilots. There is not a single word in the thirty-seven-page report that even alludes to UAP or UFOs. The perfidious transition from concerns about UFOs impact on aviation safety to a counter-terrorism exercise is indicative of the political sensitivity of this topic.

The lessons to be learned should include that neither people nor agencies like being yelled at and will likely respond negatively if they are confronted with unattractive options. That is especially true if you expect them to be of assistance in the future. Also, from the perspective of any elected officials, supporting the request to hold hearings on UFOs, or on similar sensitive topics, comes at significant political risk. It is imperative that everyone involved in the process, including those agreeing to listen to the hearings, establish and maintain credibility. Like virginity, once lost it is impossible to regain.

For the record, Sensenbrenner did take over the Judiciary Committee the year following our earlier attempt at hearings, thus providing the vacancy in the House Committee on Science, Space, and Technology we had hoped for.

Summary

For years there have been public outcries from the UFO community for Congress to hold hearings on their favorite topic. This is very naïve. For any politician, appearing to support UFOs could be the kiss of death, even when they employ disclaimers such as they are only responding to constituent inquiries. While there are some Members of Congress who tacitly support UFO research, they are loath to publicly embrace the issue as they will be forever stained. Contrary to the perception of the true believers, UFOs are not important to the vast majority of Americans. In short, from a political perspective, UFOs are not a voting issue and come with a significant downside. This chapter described the volatility and sensitivity that would accompany any attempt to officially raise UFOs in public fora.

The most important factor in the misunderstanding about how U.S. Government organizations regard UFOs is a basic lack of knowledge about how they really work. The following chapter describes the roles and responsibilities of departments thought to be involved in such research. In reality, the Golden Rule plays a huge part. In the end, it is guessing and wishing by the UFO community that have created many unsupported myths and conspiracies.

5

THE GOVERNMENT AND UFOS

What I want is to get done what the people desire to have done,
and the question for me is how to find that out exactly.
—ABRAHAM LINCOLN

Public Assumptions and Trust

Please don't skip this chapter. It is absolutely essential to understanding what the government knows, or doesn't know, about UFOs—and the facts are probably far different from what you currently think. This is especially true for most UFO enthusiasts, but holds for the general public as well. After all, polls have shown that a very large portion of the population believes that the U.S. Government is withholding information about UFOs. In 1997 a CNN/ *Time* poll found that 80 percent of Americans thought that their government was hiding knowledge about the existence of extraterrestrial life. In the same poll 64 percent believed that intelligent aliens have contacted humans. A short time later, in 2002, a Roper Poll reported similar results and indicated that 72 percent of their respondents believed the government was not telling what it knows about UFOs. More specifically, 68 percent believed the government was hiding information that was known about extraterrestrial life. Of the Roper respondents, 12 percent said they, or someone they know, had seen a UFO at close quarters. In addition, 2 percent indicated that those close-quarter events included contact with an ET. While 2 percent may not sound like a lot, it would mean that six million Americans believe it had happened to them or a close friend.

In 2007 MSNBC conducted a less scientific poll concerning ET visitations. Only 12 percent of the respondents indicated they did not believe it had already happened. Again a majority, 63 percent, believed ET has come, while 25 percent thought that such contact may have occurred but needed more evidence to be convinced.

Those are astonishing numbers and certainly indicative of a huge credibility problem for the government. Given the political climate that has existed for the past decade, trust in public officials is a vanishing commodity. While most of the known deception has been linked to very terrestrial incidents (WMDs in Iraq, Saddam's support of 9/11, the need to pony up hundreds of billions of dollars to bail out banks from a self-induced collapse and the failing automotive industry), the UFO issue adds a small, but not insignificant, piece to the collective opinion that our elected leadership is less than honest, and hardly operates with promised transparency.

As was mentioned in chapter 1, when coordinating the UFO study issue, members of the ATP study too believed that somebody must be minding the store. It was not until several years were spent gathering data and meeting with senior officials that the present understanding of the situation was derived. There is no intention to convince you that the government deserves your trust. Rather, commonsense issues are addressed that shed light on how the government approaches intractable problems in general, and UFOs in particular.

One of the most basic problems about how the public sees the role of the U.S. Government regarding UFOs is that they simply do not understand how it works. The unfortunate reality is that the foreign immigrants who acquire U.S. citizenship know far more about the functioning of our government than do the vast majority of those who are native born. Civics lessons are no longer a priority in our high school educational system and just do not appear in colleges or universities unless one majors in political science.

It is also important to recognize that "The Government" is not a single monolithic entity. Unfortunately, the media and most people talk about the government as if it were so simple. In reality it is made up of millions of people, all functioning at different levels. I often joke about being in the Pentagon and looking for the voice that talked to the media. The headlines frequently stated, "The Pentagon says . . ." However, the Pentagon is a building and it says nothing. People in that building do make statements but they should rightfully be attributed to the Secretary of Defense, or the person making the comment, not the building.

Understanding this concept is important to the study of UFOs. When reading the issues addressed at the beginning of this chapter, there is an unstated assumption that if knowledge of UFOs existed in an office somewhere in the vast bureaucracy, then the entire government is therefore responsible for concealing the information. Like all large institutions, communication, especially between agencies, is a generic weakness. Because one person or office acquires information, it cannot be assumed that the data are shared with others. Also it should not be assumed that every inquiry is indicative of institutional interest. The ability to track the source of computer searches has exacerbated this notion in UFO circles. Just because some individual from an official agency accessed a Web site does not mean the the government is monitoring them, or necessarily even interested in the topic. Most of the visits are just indicative of personal curiosity. In addition, it is important to distinguish between an individual's expressed opinion and those statements espoused on behalf of the organization. The latter are usually coordinated, allowing the various stakeholders, who are people with a vested interest, to have input.

Large Institutions and Change

There are certain fundamental characteristics that underlie all large institutions—the U.S. Government included. First and foremost institutions exist to serve themselves. Maintaining their status and staying in power are their chief concerns, thus making them increasingly conservative and resistant to change. Those qualities are not all bad, as frequent changes in priorities or direction lead to instability and lack of confidence. It is recognized that change is constant, but the rate varies. However, one of the conditions institutions fear most is disruptive change. That is unprogrammed change usually precipitated by a catastrophic event, such as the January 2010 earthquake in Haiti or the tsunami that inundated low-lying areas of Southeast Asia in 2004, both causing major loss of life and restructuring societies. Or disruptive change may come from rapid infusion of new technologies that dramatically alter daily life for constituents. Certainly rapid advances in information technologies have generated tremendous social change based on how information is disseminated and acted upon. Moore's Law states that information processing capability will double about every eighteen months, which has already led to faster communication on a global basis. Incidents happening in one part of the world are transmitted abroad instantly, despite attempts by some governments to stop them. *Twitter, Facebook,* and *eBay* have all demonstrated how rapidly social norms can shift in both communications and economics. This affects those interested in UFOs as reports and photos are circulated in near real time. Of course the downside is that there is no time for reasoned analysis and erroneous and fraudulent entries proliferate.

When faced with change that can be predicted, but with results that are likely to overwhelm the institution's capability to respond, a frequent strategy is to ignore the evidence as long as possible. Consider how governments have responded to ever-mounting documentation that climate changes were occurring on a global scale. The alarms have been ringing for decades but action has been limited. Ignoring the problem can come in the form of studies. Administrators faced with monumental issues always resort to calling for another study. They claim there is not enough data to support a response at that time and can always find scientists with a contrary opinion. And, that is what happens when highly visible issues are put on the table.

While global climate change is a series of high-probability, high-impact, and resource-overwhelming events, UFO observations fall into a totally different category. To date they have exhibited low-probability, low-impact characteristics—except for the few individuals who state that their life has been disrupted. That is, while a limited number of verifiable events have been recorded, they have had little disruptive impact on societies. The probability of an all-out invasion by hostile extraterrestrials may rank high in Hollywood, but it is deemed as an extraordinarily low probability by officials who would be responsible for organizing a human response. It is in this context that the role of the U.S. Government's involvement in studying UFOs should be examined.

How Things Work—It's Not What Most Think

Of the three branches of the U.S. Federal Government, the Executive Branch is by far the most important when it comes to UFOs. It has the organizations that do things. The Legislative Branch has some budgetary and oversight responsibilities, while the Judiciary Branch is pretty far removed from the action. While we are the most litigious society in the world, Federal lawsuits about UFOs have played only an extremely minor role when it comes to advancing knowledge of the field. Lawsuits to facilitate Freedom of Information Act (FOIA) requests did produce a small amount of information. In the case known as Cash-Landrum—which will be covered in more detail later—the legal actions against the U.S. Government were thrown out as a causal relationship between the injuries sustained by the individuals and the actions of any government agency could not be established. Such legal interventions are quite rare.

The public in general and UFO buffs in particular assume that the government must have an interest in UFOs just because these anomalies are reported. Unfortunately that is not true. Large institutions are very bureaucratic in nature. In the Executive Branch each agency has specified functions. These functions have been legislatively delegated and can be found in the *Code of Federal Regulations* (CFR). There are fifty subdivisions to the CFR and they describe the broad functions and responsibilities of all of the agencies of the government. You will frequently hear senior executives explaining their actions in terms of the duties and responsibilities under the titles that apply to their organization and that establish their legal responsibilities.

An underlying assumption for most people is that responsibility for UFOs would belong to the Department of Defense (DoD). That is a reasonable initial assumption. As we saw in chapter 3, all of the early studies focused specifically on determining whether or not UFOs constituted a threat. Under *Title 10, Armed Forces* of the CFR, the Defense Department is tasked with maintaining the national security of the United States.

Of course, under these titles the Department of Defense is given broad latitude in determining what actions are necessary to accomplish the functions of national security. However, the Department of Defense is requirements driven. That means that everything that is done is based on validated written requirements. For most functions establishing requirements is a rigid and extensive formal process. As an example, when nonlethal weapons were first being seriously considered in DoD, there were no written requirements. Even though there was an urgent need to field such systems, it literally took years to get the first formal requirements coordinated and approved.

The Budget and Priorities

As big as the DoD budget is, it remains the aforementioned zero-sum game, meaning for every project that gains funding, another must lose theirs. Interdepartmental, as well as intradepartmental, battles over funding are vicious and played by the most ruthless, merciless bastards on the planet. The DoD budget process is driven by what is known as the Program Objective Memorandum or POM. The budget cycle addresses money being spent in the current year and for funding into the future for a total of seven years. This is a terribly complex process and details are not necessary at this point. What is important to know is that it is a continual process in which programs are constantly being evaluated with a myriad of comptroller vultures waiting to pounce on any perceived programmatic weakness. This has been particularly true in the past decade as massive funds had to be allocated to fight the Global War on Terror (GWOT), including invasions of both Afghanistan and Iraq. Most readers will probably not be familiar with the budgetary impact of GWOT, under President Obama called Overseas Contingency Operations, as it has affected the military services—it has been devastating. Just note that most research and development programs, as well as training funds, have taken tremendous cuts and are constantly under scrutiny as bill payers (sources from which funds can be transferred to other projects).

One of the most mystifying terms used in the Pentagon is: "I'm going to find the money." When executives say they are looking for money, they really mean they are going after funds that are already dedicated to some current project. Understanding this process is important only in that provides the reader some context about how a UFO project might get funded. The point here is where are the requirements for DoD to be involved in studying UFOs? Experience says they don't exist.

Contrary to popular belief, there are no black projects that are exempt from oversight. This is a major misconception promulgated in the civilian sector by people who have no understanding of what is referred to as the *Black World*. It is true that there are highly classified programs that receive less scrutiny than other, open source programs. However, within those domains there are technical review committees and bodies that are intensely interested in what gets funded and what does not. Just like in the *White World*, as those open source programs are known, the competition for money, people, facilities, etc., remains fierce. Here too it is a zero-sum game, and the same issues of diminishing resources come into play. While the Legislative Branch of government, Congress, is responsible for providing funds to the Department of Defense, they are also charged with oversight. With highly classified programs there are fewer legislators involved in the process, but they are there.

There Is No Alternate Chain of Command

An important point that rebuts the conspiracy theory notion that people are required to lie to senior officials about the existence of a UFO program is an apparent lack of understanding of where the DoD fits under the Executive Branch. The whole thing, including all of the agencies, comes together under the President of the United States who is also the Commander-in-Chief of the Armed Forces. By law, the Secretary of Defense reports to the President. By law, the officers and civilian appointees of the DoD report to the Secretary of Defense. By law, there is no place where one is allowed to deny information to those positions. This notion of legally withholding information from these key people is a fantasy that has been proffered by theorists who have to explain why certain officials have no knowledge of the UFO topic. Secrecy and how it impacts the field of UFOs is covered in Chapter 12. In understanding how the Executive Branch works, note that the President is in charge—over everything. Just because a program is black does not allow underlings to determine the validity of reporting required information.

From a practical standpoint, that does not mean the President knows everything that is going on in his or her administration. One of the biggest problems at the highest senior leadership levels is determining what information is important and providing data in a highly synthesized, yet understandable, manner. There are simply too many balls in the air at any given moment for any single individual to have cognizance of them all. The senior executives have to know which issues are rubber and will bounce if dropped, and which are made of crystal and likely to shatter on impact. Unfortunately for people interested in phenomenology, at the POTUS (an acronym for President of the United States) or White House levels, UFOs do not fit in either category.

There are many organizations not related to the DoD that are responsible for extensive research on a broad range of topics. In fact, the national laboratory system is conducted by the Department of Energy. And there are smaller institutions, such as the National Science Foundation, that also fund research projects. However, there is no *Department of Good Ideas* that is tasked with looking into anomalies, no matter how important they might appear to some people. In order for any topic to be researched, it must fit into some bureaucratic box so that the funding can be established. Americans demonstrate an extremely high degree of ingenuity. They are constantly coming up with good ideas that are worthy of serious consideration. From a government funding standpoint, the problem for people with those good ideas becomes finding an organization with compatible goals and responsibilities. It is also important that the agency has the funds available and that the concept is adjudicated at a level high enough to obtain support.

The real funding issue is that all of the U.S. Government agencies run out of money long before they run out of good ideas and approved projects that deserve attention. All agencies have "1-to-n" priority lists. The most important project is assigned 1, and the list runs in descending order to "n," meaning some unspecified number that is last on that list. Included on the lower end of these lists are what are called unfunded requirements. These are approved projects with established requirements, but there's not enough money allocated in the budget to cover them. The problem of funding a UFO project can be understood if the readers could place themselves in the role of a senior manager. They would be encumbered with legal, fiscal responsibilities; ones that can come with penalties including the possibilities of fines or incarceration if violated. Then one should determine where support for UFOs would fit when compared to all of the other priorities duly assigned to that organization. At least in theory, the decisions should be made based on defensible logic, not emotion. Again, in all government budget priority lists, the organization runs out of money long before they run out of projects with validated requirements. This is the environment in which a UFO project would have to compete for funding and other scarce resources such as qualified personnel. When viewed from a real-world management perspective, it is hard to envision a senior official risking other projects of their agency, or in extreme cases their career, just to fund a UFO project.

How Bureaucrats Think

Bureaucrats tend to think like engineers or bricklayers. The applicable motto could be "A place for everything, and everything in its place." In other words, they are very concerned about which box to put things in, but often couldn't care less about the outcomes of specific actions. While a bit of an overstatement—as there are certainly many professional people who take pride in their work—too frequently senior bureaucrats are more concerned about organizational propriety than the product. Remember the lesson from the Condon Report. Regarding that study, the desired outcome for the U.S. Air Force was to be relieved of responsibilities for investigating UFOs. Mission accomplished.

For many bureaucrats, issues and actions that cross institutional lines are perplexing, especially if they have a potential for undesirable characteristics or consequences. Like UFOs, whenever possible such issues are to be pushed on to some other organization, or simply avoided entirely, with the hope that some event won't come back to bite you. While usually small, there are huge cross-jurisdictional issues, such as those leading up to 9/11, that have catastrophic results. Simply put, avoid tar babies.

Worth noting is this bureaucratic infighting is exactly what happens with UFO sightings. They pop up from time to time, as was demonstrated when Congressman Steve Schiff of New Mexico asked the Department of Defense for an explanation of the Roswell event. First he was ignored. Then he was given an incomplete and dismissive response that did not address the facts, which was signed by a relatively lower level staff officer. Schiff followed up by addressing the Secretary of Defense directly but still never received a fully satisfactory answer. Following classic bureaucratic maneuvering, the action was passed to the General Accounting Office (GAO) for processing. In the end the Air Force was plagued by some minor, institutionally irritating press, and the topic drifted off into oblivion. For those interested, the GAO report can be accessed on the Internet. In short, the GAO report stated they contacted all of the usual suspects (DoD, CIA, FBI, etc.). As a result, they acknowledged that something unusual happened on July 1947 near Roswell, but that it was not a UFO crash.

The well-known advice from *Deep Throat* to Woodward and Bernstein—to "follow the money"—remains valid, and the funding process of Congress is worth considering when attempting to understand the workings of any program, including ones that might involve UFOs. By design there are two distinctly different congressional bills involved in providing funds to all governmental agencies. One part of that process is the Appropriations Bill. It is this bill that establishes the amount of money that can be used by the agencies and for what purposes and sets it aside in the U.S. Treasury. Then there is the Authorizations Bill, which becomes the public law that authorizes the agencies to actually spend the money for the projects that are addressed in the bill. To spend money on programs not specifically approved is an illegal act.

Obviously some funding can be, and is, hidden. The public has been made aware of some of these misapplications of funds. That is why specific funding thresholds are established throughout the process. Circumvention of the process, such as happened in the Iran-Contra scenario, is illegal. The result was demonstrated when senior Reagan officials were charged with Federal crimes. At lower levels of the organizational structure the amount of money that can be reprogrammed from one project to another is relatively small. As the amount of money to be moved increases, so does the level of authority for reprogramming. For large sums to be transferred, agencies are required to request congressional approval before the money moves. Failure to follow these rules is a crime.

There Are Many Players, and Not All Are Visible

How the appropriations or the authorizations bill gets written is intentionally mystifying to the public. This is where the lobbyists get actively involved. There are many offers to assist the Congressional staffers, who are generally well meaning, but constantly overworked. Of course the lobbyists generally have a dog in the fight, meaning a vested interest in the outcome. Therefore their assistance is self-serving and should be viewed with extreme caution. Rampant examples of lobbyist influence were apparent in congressional actions regarding health care, banking, and finance. For perspective, during the health care debates there were an estimated six lobbyists for every Member of Congress. Harvard law professor Elizabeth Warren, the Chairperson of the Congressional Oversight Panel (formerly known as TARP), noted that banking lobbyists hit every Congressional office four to five times per day. Strong, too, in pleading their cases have been the tobacco lobby and the NRA. The legislative results are obvious.

For comparison, there is only one lone registered UFO lobbyist. For many years Steve Bassett has been making the rounds, talking to whomever he can about issues of disclosure of whatever the government knows about the topic. Process-savvy, Bassett has been able to energize people interested in UFOs to make noise and at least let staffers know that someone is concerned. As a one-man band you can imagine his ability to influence policy makers as compared to the powerful special interest groups just mentioned.

One purpose in addressing Congressional funding is to point out the number of places that oversight takes place. Both Senators and Members of the House of Representatives covet positions related to appropriations and authorization. They are positions of power for both them and their staff members. In order to get money approved, information must be provided to these committees. Even the Black World budget has some people looking closely into programs. Access to those classified programs also yields a higher degree of power. Remember, the committees mentioned are chaired by a member of the dominant party in Congress. The leadership has changed hands many times during the period since UFOs became an issue. In a city that runs on power and influence, I'm confident that if a UFO program was being funded, especially over a long period of time, the word would have come out through official channels. The Golden Rule is preeminent, *"He who has the gold makes the rules."*

In recent decades a new breed of power brokers has emerged. Janine Wedel, a professor at George Mason University, in her recent book *Shadow Elite*, called them flexians. These flexians move seamlessly from opposing positions of power, from key civilian enterprise into the government bureaucracy and back again. They make the rules, then bend them as they create new organizational structures with personal loyalties never quite assured. Flexible by nature, hence the name; it is often hard to pin down who they really are or what they represent at any given time. Among these flexians are the princes of Wall Street, who then become Treasury secretaries, or they are on the Security and Exchange Commission or enter other positions that control national and international finance. The flexians have re-created a world with the *Inverse Golden Rule, "He who makes the rules gets the gold."* If UFOs, with the potential for wealth that they bring, were reposed either inside or outside the government, the flexians would have taken control and capitalized on it. It is they and their colleagues who populate critical legislative and executive administrative positions before moving on to other sectors of the economy. Most important, they always maintain access to the pulse of knowledge and remain positioned to influence the action. When you consider the number of people who would have had access to critical information over a period of more than half a century, the resulting silence on UFO matters does not make sense.

Personal Interest Versus Institutional Responsibility

Another public misperception is confusing expressions of personal interest in any given topic with those of institutional interest and responsibility. This means that just because certain key individuals express a personal interest in UFOs does not mean that their organization accepts responsibility or is involved in studying the phenomena. For organizations in the Executive Branch, senior administrators are bound by laws that describe their roles and responsibilities and restrict what they can and cannot fund. At times conflict emerges within the organization when a senior official interjects his or her personal beliefs—often religious in nature—into administrative decisions such as funding or wording of reports. In recent years such conflicts have become apparent in fields such as health care and the environment.

However, when these conflicts occur, they often make it into the press as the B team leaks the information to news sources that abound and eagerly await any hint of malfeasance. The B team refers to the myriad of career civil servants who make all governmental institutions function. As they sometimes say—rarely overtly—to leaders attempting radical change, "I be (B) here when you come, and I be here when you leave." Leaking information is a classic tactic for impeding change for things that they do not approve.

Government Employees See UFOs Too

Just like the general population, a known percentage of government employees have had personal observations of what they believe to be UFOs. That constitutes a personal interest. Sometimes they will be in positions to scan files and see what information is held in official files. That still does not mean their organization has an official interest.

The best example of personal versus institutional interest previously mentioned was with the former head of one of the lettered intelligence agencies. He noted that his agency did not have any requirements to collect information about UFOs or to study the topic, despite his personal observation of UFOs that had taken place long before he had assumed the directorship of that agency. His personal conviction that UFOs were real was insufficient to translate into any formal collection requirements for his agency.

While he was the highest ranking official who provided me a firsthand observation, the experience was not unique. Once people became aware of my interest in the topic, it was not uncommon for them to relate personal experiences. As an example, one day while at CIA Headquarters in Langley, Virginia, on other business, an introduction was made to several agents who wanted to hear about UFOs. Once they seemed comfortable that this topic was taken seriously, two of them described their personal observations. One of those was a daylight sighting in a rural setting at relatively close range. The woman and her partner were driving along when the object appeared over a farmer's field. It had no wings and did not make any sound that they were able to hear. The incident went unreported as it was not germane to the mission. While it obviously made a mental impression on them, it did not alter their business agenda and does not infer that the CIA as an organization has an official interest.

There are thousands of similar examples in which individuals have had personal sightings of UFOs, yet that have not impacted their work performance. The operative statement would be "Not my responsibility." It is known that many military pilots have seen UFOs and have made a conscious decision to not go on record, generally for fear of ridicule. Reporting UFO sightings is not career enhancing in any government agency.

It is also a mistake to assume that every inquiry about UFOs made by someone who happens to work in a government position is an indication of official institutional interest. Unfortunately, in several documentaries there are such implications made. The assumption appears to be that if a report goes to an agency through official channels, therefore that organization must be conducting a study of that subject. This has been used in many cases to imply that the U.S. Air Force did not stop investigating UFOs at the end of Blue Book. Remember, there are millions of individuals working for the U.S. Government and they come with a wide variety of beliefs and experiences. Some of them have had personal observations. Like the general public, many government workers are curious about the topic. Contrary to the documentary comments, personal interest or inquiries, or even incidental reports, does not constitute the government officially studying the topic. Far too much is made of such allegations.

Interestingly, official responsibility is not even understood by some of our senior leaders. The case known as the Phoenix Lights will be covered later. It was a major incident and eventually caused Arizona Senator John McCain to make official inquiries. He seemed surprised that both the Air Force and the National Archives stated they don't investigate UFOs or keep those files. Here the Air Force could have been more prudent in its response, but it frequently insists on self-inflicted wounds rather than being polite and understanding the seriousness with which many people take this subject.

Playing to public sentiment, on a radio program on June 25, 1997, McCain stated that "people saw things and whenever that happens they deserve at least an investigation." Really? There are many unusual observations that are made every year. What criteria should be used to determine whether or not an investigation needs to be made? As mentioned, the government is a huge bureaucracy and investigations happen for specific purposes. At the moment, UFOs do not have such a requirement. Condon removed that responsibility for them as the Air Force investigates possible threats.

Everything Leaks in Washington

In Washington information is a precious commodity. Especially at the policy level, where knowledge is power, the ability for one-upsmanship is widely employed as a means of getting what one wants, so keeping a secret like UFOs would be difficult. One has only to consider what has leaked in that city to understand how intertwined information and power are. From major scandals such as Watergate to insignificant sexual peccadilloes like President Clinton's escapade with Monica Lewinsky, within a relatively short time details come spewing forth—often more than you'd care to know. For political purposes Valerie Plame's position at the CIA was revealed. The warrantless eavesdropping of the NSA became headlines, even though countering terrorism was a critical national priority and information about the project was extremely sensitive.

There is no doubt that inappropriate events occur, but they are usually revealed, often by a whistle-blower like Fred Whitehurst who demonstrated that the famed FBI laboratory could not distinguish bombs from urine. It was Whitehurst whose revelations rocked the foundation of the top forensic laboratory in the country by complaining about shoddy analysis that was sending people to jail. During the Vietnam War *The Pentagon Papers* were leaked. In the civilian sector, Jeffrey Wigand informed us that the tobacco industry was manipulating the nicotine content of their product and Sherron Watkins became infamous in the demise of Enron.

Added to this volatile mixture are the ever-present reporters who are out to become the next Bob Woodward or Carl Bernstein. Many are inexperienced and recklessly blow stories out of proportion. Today uncontrolled bloggers compete with mainstream media to break stories and sometimes engage in questionable procedures. In this age of 24/7 news coverage ethical reporting has become an oxymoron—speed is far more important than accuracy.

Some Secrets Have Held—For a While

It is true that there have been some secrets that have been held over time. Usually, however, some information about the project gets leaked. One of the best examples would be *Have Blue,* the predecessor to the stealth aircraft. Have Blue was a DARPA project, executed by Ben Rich at the Lockheed Skunk Works. It began in 1976 and the first flight was during December 1977. That ended and a follow-on program, *Senior Trend,* began. Those programs eventually led to the F-117, which is still erroneously called a stealth fighter, and first operational capability was achieved in October 1983.

Providing the extreme security for the development of the F-117 added at least 15 percent to the overall cost of the project, but it was well worth it. However, flying anything for a long period of time is hard to conceal, even in the vast Nevada deserts. Before long people outside the program were aware that something was in the mill, even if they didn't know the details. In fact, by 1986, before the F-117 was revealed, Testors, a model company, released a kit called the F-19 based on what they believed was a stealth fighter. Revell joined suit and sold their F-19 model as well.

As one of the most important technological advances in aviation, and with a critical war-winning mission, some information gradually came out and the aircraft was formally declassified when it was used in support of Just Cause, the invasion of Panama. It was on August 1, 2008, just more than thirty years after conceptualization, that the aircraft was retired. The point is that this ultrasecret project went through its entire life cycle in a relatively short period of time. Yet many Americans believe that a project of more interest could remain hidden for so much longer.

The site where the nation's secret aircraft was developed and tested, Area 51, has become almost mythical, even as the Government was foolishly denying its existence. Area 51 was incorporated into movies such as *Independence Day,* as well as a host of television productions and computer games. With satellite photos of it available on the Internet it was among the worst-kept secrets in the Department of Defense. Finally, in 2008, the veil was lifted, a bit, and the people who had worked so diligently to protect the nation were allowed to speak publicly for the first time. They were even allowed to call the site Area 51. Those interested can locate many of the stories through their organization called *Roadrunners Internationale.* Incidentally, several of these members have spoken publicly and discussed the possibility of UFOs being housed, or even flown, at Area 51. While not putting stock in the rumors, they are fully aware of the folklore attendant to their site and find it rather amusing.

Of course there have been a number of whistle-blowers who have come forward regarding UFO tales. Some of them have become relatively famous in the process, at least within UFO circles. The difference between the UFO whistle-blowers and those previously mentioned is that when scrutinized their stories usually fall apart. Few of them turn out to be real, in the sense that they worked where they claimed they did, or had the access they stated. One of the most infamous of those was Bob Lazar, who claimed to have worked at S-4, a secret subterranean element next to Area 51 near Groom Lake. While there he claims he was given access to many technical documents about UFOs and saw the craft in the hangars. The problem comes when Lazar, and others like him, get caught in the details. For Lazar these include bogus claims of advanced degrees from Cal Tech and MIT. Neither institution has a record of him ever attending, but his mediocre academic achievements were located at a junior college in California.

According to Lazar it was Dr. Teller, whom he simply calls "Ed," that arranged the job for him at Area 51. For the record, I asked Dr. Teller about this and he indicated he had no knowledge of Lazar. Despite possessing minimal technical skills, we are led to believe that his expertise was so important that he instantly bypassed the months to years of background investigations to which we mortals are subjected. Despite overwhelming evidence against him personally and his tale in general, his story was so popular that Revell Toy Company created and sold a UFO model that bore his name. Now there are even rumors that Matt Damon may play Lazar in the movie version of this fairy tale.

Real secrets almost always get revealed—often before they should. In Washington, information manipulation is a fundamental attribute of the game called politics. Awareness of actual contact with extraterrestrials would be a trump card, one irresistible to play at a strategic moment. To think that amoral politicians would withhold the information due to some altruistic sense of loyalty, or just because the UFO topic is so special, is ludicrous.

So Where Would It Be?

As indicated there are several institutions that might have legitimate interest in UFOs. We have explored the most popular one, the Department of Defense. While many will disagree with this statement, there is no big black UFO program there.

The Department of Energy's Role

The next best bet would be the Department of Energy, as it controls all of the national laboratories and has immense scientific capabilities. Considerable experience at Los Alamos National Laboratory (LANL) suggests that DoE does not have such a program either. What was found while working at

LANL very much paralleled the experience in DoD. Again, there were people with both interest and personal observations, but no program. Some of the better-known photographs were made available to interested scientists at LANL as they had access to advanced computing power and photoanalysis capabilities. In all of the known cases, this analysis was done based on the personal interest of the researcher involved.

Some of my friends were among the best and brightest scientists in the world and had been employed at LANL since the early days. These guys knew the original giants of nuclear science personally. Some had even discussed the topic of extraterrestrials with legendary figures such as Italian-born Enrico Fermi. In 1938, at age 37, Fermi received the Nobel Prize in Physics and was later known for the development of the first nuclear reactor. Because of the anti-Semitic edicts of Benito Mussolini he immigrated to the Unites States immediately after accepting that award and was instrumental in the Manhattan Project that developed the atomic bomb. For his extensive contribution in nuclear physics, *Time* magazine named him one of the top twenty scientists and thinkers of the twentieth century.

One of my friends from LANL, who chooses to remains out of the public eye on matters such as UFOs, was present when Enrico Fermi asked his salient question, "So where are they?" He went on to hold the Fermi chair at the lab. That means the chair that Enrico Fermi actually sat in while he worked in the office next door. Fermi, like Sagan after him, had noted the billions and billions of stars, and based on probabilities, believed we should have been contacted by intelligent ETs. That we had not been contacted became known as the Fermi paradox.

The problem that the Fermi paradox raises for those who suggest that we have made contact, or may have recovered alien bodies, is that these most illustrious scientists had no knowledge of any of the interactions claimed by conspiracy theorists. Surely, if we had been in contact with ET, or made the recoveries suggested by Roswell devotees, our most respected scientists like Enrico Fermi and Edward Teller would have been involved and the word would have spread more widely.

Los Alamos and UFOs

Rumors abound regarding underground chambers at Los Alamos that are used to house ET, or store UFO material. In fact, some subterranean facilities do exist there. Many of them are used for storage of nuclear materials—which by their very nature are subject to extremely high security measures. I had two incidents related to claims that these sites were used for ET material. One came as a result of contacts with Bill Moore who is known in UFO circles for exposure of the Majestic 12 documents. These documents will be covered in more detail in Chapter 7. The issue here is specific to underground facilities at LANL. Moore set up a meeting in Gallup, New Mexico. Attending were physicist Hal Puthoff, Congressional aide Scott Jones, and myself. We were to meet with a source who claimed to have been involved in an official UFO project when he was in the U.S. Air Force. While the identities of Puthoff and Jones were known to Moore, I was an unknown. He guessed, incorrectly, that I was an officer from the Defense Intelligence Agency (DIA). In reality I had retired from active duty and was already working at Los Alamos. As I told Moore at the time, I was not with DIA nor had I ever been assigned there.

The "source" gave each of us a rather convoluted story about ET and his involvement in the project. He went into substantial detail about his induction into the program that had taken place at an underground facility at Los Alamos. The numbers of people he suggested were involved in the program were hard to believe. It would have been nearly impossible to keep a secret of that magnitude under wraps for half a century. More important, he described a facility that I knew pretty well. Having recently been in the building he specified, I knew that there were no underground bunkers in that facility. This source also did not know that recently this technical area had been opened for public access. As a cost-cutting measure at the end of the Cold War, many technical areas that had previously been guarded were opened to the public. The fences he described remained in place, but nobody manned the gate. Any civilian visitor could have approached and entered that building. That hardly described a facility that holds the nation's crown jewels.

Years later a second incident came in an evening phone call from Bob Collins, a retired USAF officer who did know my identity. He also knew that I was working at Los Alamos, but not the exact location of my office. At the time retired Air Force Colonel Jerry Perrizo was the leader of a group identified as IT 6. The nature of the work of that group was very sensitive. It has been publicly identified by LANL as an intelligence technology group but in no way were its efforts related to UFOs. Captain Collins informed me that he thought the underground facilities were located in a remote site known as TA-33. The ironic thing was that I was sitting in my office at TA-33 and was quite familiar with the complex. It is a fairly large technical area and it is restricted from public access. At lunch I routinely went running and had covered most of the complex on foot. There were a few areas that were prohibited to all personnel. That was not because ET artifacts were stored there, but because the areas remained radioactive from some of the previous experiments conducted as early as the Manhattan Project. It was due to the potential exposure to radiation that all workers at the site always wore dosimeter badges that were collected and analyzed monthly. Anyone interested in checking Collins's claims can access his material that is published on the Internet. Maps clearly identifying TA-33 can also be located via Google Earth. There were some very interesting projects conducted at TA-33, but none of them involved UFOs.

The FAA Does Get Reports—But Sends Them to a Civilian Organization

The Federal Aviation Administration (FAA) was another likely candidate for knowledge of UFOs. After all it does monitor American airspace and has been called by a substantial number of people wanting to report their unexplained observations. It is also a favorite watering hole for UFO researchers who want to validate reports. For the most part, the FAA reports that it did not have contact with the objects in question. But there are surprises.

To clear up one general misperception, the air traffic control radar system works quite differently than military radar systems. For civilian airlines, the system is cooperative. That means that each aircraft has a transponder that is actively sending out a signal. These pilots want to be seen and for the controllers to know their status at all times. The military radars send out signals that bounce off aircraft and the return is measured. Since UFOs operating in civilian airspace are not part of the normal flow of traffic, it is not surprising that they are not tracked by the FAA.

I had personal experience with this agency, albeit via synchronicity. When first assigned to the Pentagon I lived in an apartment complex called Alexandria Knolls West, which was located just off Shirley Highway. One of the people living there was a fine gentleman, Quint Johnson, and he happened to work for the FAA. In fact, at that time Johnson was the deputy to the director of security for the agency. We maintained contact over the years, even after I left the Washington area. Our primary point of discussion regarding work revolved around the lack of security at airports, especially those in foreign countries. Johnson lamented the poor state of affairs, and his office had tried in vain to beef things up. Their primary opponents, believe it or not, were the commercial airline companies. They lobbied hard against any procedures that might raise concerns with passengers. Their emphasis was on moving people as quickly as possible, minimizing contact time between agents and people, and telling them to have a nice flight. Despite years of unheeded efforts to improve airline security, one week after 9/11 he and his boss were designated sacrificial lambs and pushed out of their jobs.

In the mid-1990s I moved to Las Vegas and was working for Bob Bigelow at the National Institute for Discovery Science (NIDS). Bigelow created NIDS to explore two topics: the continuation of consciousness beyond physical death and UFOs. A reported billionaire, Bigelow was in a position to put substantial resources into this research. One of his concerns was the lack of current cases and establishing a process whereby NIDS might gain rapid access when cases were reported. We decided to approach the FAA, as they were a logical agency that was highly likely to be involved in receiving UFO reports.

The first call was to Johnson to arrange a meeting at his D.C. office on Independence Avenue. Johnson was aware of my background in studying phenomenology, but was highly skeptical that his agency might be willing to be involved. He had no information about prior involvement of the FAA, which would indicate that they did not view UFOs as a security matter. Johnson did provide us entrée into the FAA at a reasonably high supervisory level and meetings were arranged.

The key meeting was attended by Bigelow, Colm Kelleher, the Deputy Administrator of NIDS, and me, plus several supervisory personnel from the FAA. Somewhat surprisingly, these FAA representatives were open to discussing the topic. It helped that one of the senior people had come up through the ranks as an air traffic controller. He acknowledged that while in those positions he had encountered reports of UFOs. However, there was no central collection point for those reports and no secret agency to which they had to send the information. Basically, when UFOs were reported nothing happened unless there were special circumstances such as the press asking questions. In assuring the FAA that NIDS was volunteering to be their 911, and that they would not assume any risk or cost, they agreed to assign the responsibility to NIDS and did post the information in their operations manual. After NIDS was closed, Bigelow established a follow-on organization called Bigelow Aerospace Advanced Space Studies (BAASS) to pick up the mission. It is important to note that interest in the UFO topic was initiated by a civilian company, not the FAA. Like the Air Force, they were happy to have the matter go away.

Maybe Defense Contractors Have It

A popular proposal among conspiracy theorists is that the UFO program was moved either totally outside government control or into some hybrid consortium. According to this theory, at some point in time the decision was made to privatize further development and the recipients were one or more of the large defense contractors.

If one based this rationale only on developments within the past two decades, there would be reason for concern. As indicated, the premature jubilation and expenditure of the peace bonus yielded a dramatic reduction in the size of many government agencies, especially the Department of Defense. With the advent of GWOT, it was quickly apparent that the force structure was totally inadequate to handle the resulting campaigns. The answer was to turn to contractors in numbers previously unheard of. The ratio of military forces to contractors was about one to one. For the surge of 2010 in Afghanistan, more than 60 percent of the force structure was contract labor.

However, it was not just in foreign theaters that contractors prevailed. In almost every top military headquarters there were contractors in sensitive support roles, some reporting directly to the commanders. While there have been contractors in all of our recent wars, never have there been so many holding key positions. In wars past, contractors were brought in as technical representatives to assist in maintenance of the advanced weapons systems that had been bought from the large firms. In GWOT, contractors were everywhere, including being involved in armed engagements.

While budget decisions are left to official government workers, much of the staff work that goes into the development of the POM is done by contractors. This is not to question the loyalty of these people. Many of them held significant positions in the military before they retired and transitioned to defense contractors. For disclosure, I also joined those ranks and went to Afghanistan in 2003. The task was to be an advisor to the senior leadership of the new Afghan National Army. This job should have been assigned to some active duty colonel or general, but the DoD was totally committed, even at that early date.

The point for UFO-watchers is that it would be easier for contractors to move money to such a project given their access to the finance cycle. The reality is that given the funding necessary to conduct GWOT, the resources available for diversion to any project, let alone one regarding UFOs, have been diminished dramatically.

However, there is another problem with the defense contractor/UFO theory—what can be called the "begats." Industrial downsizing began even before the end of the Cold War. Not only did the personnel system take a beating, so did the requirements to purchase war materiel. In short, there were too many defense contractors competing for an ever-shrinking budget. As a result, major contractors sold off units of their companies, often to previous competitors. Mergers were rampant and it was even suggested that in the end there might only be one company left standing. Within DoD there was serious concern that the cuts were so deep that there might not be adequate competition to maintain lower costs. From a worker standpoint, when you asked a person in the defense industry, he would cite the lineage as his segment of the organization was moved from one company to another, much like the way heritage is described in the Bible.

The mergers included huge companies, as well as acquisition and divestment of subsidiaries. As examples, in 1994 and 1995 Lockheed Corporation merged with Martin Marietta. Northrop merged with Grumman. Boeing merged with McDonnell Douglas. Raytheon merged with E-Systems and acquired Hughes, which had already been bought by General Motors. Previously each of these corporations had played a significant role in aerospace development. Of course, part of the merger process is full disclosure of company assets and business records. But it doesn't stop at our shores. Some of these elements were purchased by foreign-owned companies, which raises another specter—that the technology might be sold to an offshore corporation.

The question to be answered by the conspiracy theorists who would have us believe in a UFO program, with potentially revolutionary technology, is, "How could it remain hidden during the intense investigative processes leading up to a merger?" Further, the notion that a consortium comprised of fierce competitors would somehow place a UFO development project outside of their feeding frenzy seems preposterous. It should also be remembered that any U.S. defense contractor is bound by the laws of this country. If there were work ongoing for any government agency relative to UFOs, they would be responsible to that agency, and ultimately to the President. Of course there exist rumors that suggest some übersecret, extrajudicial consortium is in control of the UFO process or technology. That fails the test of common sense and is simply not believable.

Compared to What?

At the end of the day, when examining the government's role in UFOs, one must consider the ultimate issue: Compared to what? Where would a topic such as UFOs fit in the hierarchy of issues competing for attention? UFO enthusiasts would rank the matter relatively high. Most of the world would not. From the government's perspective, funding a UFO program at any amount sufficiently substantial to draw oversight would be extremely hard to justify. Yet conspiracy theorists suggest that is exactly what has happened.

It's All About Priorities

At every level of government, administrators would have to compare the importance of UFO studies against everything else for which they are responsible. At the national level top issues would include the economy, jobs, health care, ongoing wars, and similar macroscale issues. As noted, for the past two decades the Department of Defense has been under severe budgetary constraints. First they had to contend with the drawdown of force structure based on the hypothesized peace dividend—the fantasy notion that the end of the Cold War meant the end of threats to our security. Then they had to pay for the Global War on Terror, conducted with an underresourced military, and one that has almost eviscerated the machines of war. Vehicles and aircraft that were never designed for fighting in an inhospitable environment for years on end have been nearly worn out and require replacement at a cost of many billions of dollars.

On the personnel side, our troops from all of the services have been committed in combat at a rate unmatched since World War II. In fact, we now have soldiers and marines who have been in combat longer than anyone in that war. This holds true for the reserve components as well. Members of the Army Reserve, National Guard, and Marine Corps Reserve have been deployed to Iraq and Afghanistan at rates never anticipated.

Those who believe that the DoD has some active program on UFOs must answer where the resources would be coming from to execute it. Conspiracy theorists would have us believe that there are units on standby just waiting to respond to any UFO incident, especially a crash. According to their version of history, this has happened on several occasions, yet the reactions described again fail considerations of rationality.

Many UFO investigators point to JANAP-146 as proof that the U.S. Government still requires sightings to be reported. This notion comes from U.S. Air Force Regulation No. 55-88, signed into effect on May 13, 1966, and has never been rescinded. The regulation is entitled *Communications Instructions for Reporting Vital Intelligence Sightings* (CIRVIS). There is a very important statement included in this regulation that frequently is ignored by those who purport that a secret program exists. The text states, "Air Force personnel will report by rapid communication procedures all unidentifiable, suspicious, or hostile land, air, or seaborne traffic which—because of its nature, course, or actions—*must be considered a threat to the security of the US or Canada*" (emphasis added). This requirement is commensurate with the findings of the Condon Report. Since the U.S. Air Force explicitly excludes UFOs as a threat, this regulation does not apply. Contrary to the popular belief, there is no requirement for the Air Force, or commercial pilots, to report UFO sightings.

Dead Alien Recovery Example

One oft-told case stands out as an example worth considering. According to legend, on January 18, 1978, an excitable military policeman shot and killed an alien intruder at Fort Dix, New Jersey. This Army base was contiguous to McGuire Air Force Base. The shooting took place "in the early morning hours," yet according to the reports, a C-141 transport aircraft arrived at McGuire from Wright-Patterson Air Force Base at 7:00 A.M. This C-141 carried a special unit noticeable by their blue berets, and previously unseen before. They brought with them a specialized container that was used to carry the alien's body back to Wright-Patterson. The alien was described as non-human, small, slender, and fitting the description of others found at previous crash sites. As is often reported, the observers were interrogated and threatened with dire consequences should they ever discuss the incident.

Of course the problems with the case are epidemic. First there is a shooting of something (alien or human) that went unreported in official Fort Dix military police or command channels. Next there is the recovery issue. According to the reports, a plane was sent from Wright-Patterson to McGuire Air Force Base in a matter of a few hours. The relatively short flying time is not the problem. It is the logistical planning efforts that fail the test of reason. This scenario depends on a C-141 being on standby, to fly at a moment's notice to wherever a crash might occur. That means that despite the extraordinary low probability of such an event occurring, it was considered of sufficient priority to dedicate a precious C-141 cargo aircraft to that mission. Most civilians cannot imagine just what such a commitment would infer. Having participated in various Air Force war games, I can state that aviation transportation was always the long pole in the tent in every exercise. The USAF never had enough lift capability, but here we are led to believe that a large cargo plane is reserved for this improbable mission. There is also the problem of the blue beret personnel who were reported on the scene. The reference to the beret is feasible as they were approved for wear by security personnel in 1975. However, this report presupposes that some specially trained security unit is constantly available for UFO missions. The timing suggests a unit on strip alert. To man such a force 24/7 over the decades would mean that thousands of personnel had been briefed. This seems highly unlikely. For this story to be true, one must believe that senior commanders placed UFO recovery as a priority over most other missions.

The Capabilities Paradox: The Government Can't Get Anything Right—Except UFOs

Disparagement of the effectiveness of government agencies is the standard stuff of jokes. Even our legislators argue about how ineffective the government is at running programs. When opposing a bill that would have a government agency pick up responsibility from the private sector, government ineffectiveness is often used as a scare tactic. This was certainly evident in the 2009–2010 health care debate.

The fact of the matter is that the government does do some things well. The U.S. Postal Service is an example of a large organization that delivers mail service quite efficiently across the nation. Of course they are not perfect, and there are many counterexamples of ineptitude in other agencies. Recent history is replete with failures including their handling of the financial crisis that nearly destroyed the economic system. Travelers will recognize the problems associated with the Transportation Safety Administration. Health issues are raised with inadequate inspection by the Food and Drug Administration.

The capabilities paradox refers to the perceived inability of the U.S. Government to manage things efficiently, except for one distinct area—UFOs. It is this unique area in which the government is perceived as near omniscient, well organized, and highly focused. Many people believe the government can't secure our borders, especially those with Mexico. However, let a UFO fly by, let alone crash, and the responsible agencies are instantly involved in chasing the intruders or recovering material.

Possibly a better example are the mistakes made by the Intelligence Community. Historically they missed Sputnik, the fall of the Shah of Iran, the Berlin Wall collapse, 9/11, WMDs in Iraq, and can't locate Osama bin Laden. But their systems supposedly become highly capable when UFOs enter the area. Somehow these apparent incongruities again fail the test of common sense.

FOIA

The Freedom of Information Act (FOIA) was enacted in 1966 to increase transparency of government activities. While certain information was exempt from disclosure, the vast majority of reports were fair game. Among the reasonable exemptions were requests for classified material, trade secrets, and personal privacy matters. FOIA became the tool of choice for UFO investigators and all departments were flooded with requests. In fact, those requests are still coming in today. At one time it was estimated that about half of the

FOIA requests concerned UFOs, something the drafters of this law never contemplated. This abuse resulted in drastic measures being taken by several agencies.

By law, an agency has only twenty business days to respond to a request. They are to be handled in a first-in, first-out basis. The agency is required to provide records that exist, but is not expected to generate new reports. Therefore, the person making the request is expected to be clear and concise about the reports they are seeking. In addition, there are fees that can be assessed for the time and materials used to answer a FOIA request. These are usually reserved for requests that take a large amount of effort. Agencies are not required to make inquiries of other agencies that might have the information desired. It is up to the person making the request to address each agency individually.

The reality is that many agencies do not meet the required deadlines. Mostly this is from being overwhelmed with requests and having too few people allocated to responding to the public. Unfortunately, FOIA is generally not seen as a high priority in most departments. As budget cuts have sliced into personnel at all agencies, staffing FOIA offices has experienced an even lower priority. Requests for information derived from classified reports are still more difficult. These requests require security reviews that add to the response time. Routine declassification of documents, while verbally stated as a priority, is currently years behind schedule and a topic of debate by watchdog groups and Congress alike.

There have been quite a few UFO reports released over the years but many of them were highly redacted. Researcher Stanton Friedman has made quite a show over the years of displaying reports that are mostly blacked out to the point that they are useless to the reader. He has rightfully questioned this extensive use of bureaucratic censorship.

For the first years the Intelligence Community agencies attempted to respond to FOIA requests pertaining to UFOs. The volume of requests became unmanageable so a new approach was taken. They banned the intake of new UFO material. Several sources have informed me that the CIA simply put a filter on incoming messages. If UFOs were mentioned, the report was simply deleted. In that manner, the agency could state that they were not holding any UFO reports that had not been released. Today, the CIA has all of their old UFO data accessible on their Web site. You can view it at www.foia.cia.gov/ufo.asp.

Of course this is bad news for people who believe that the CIA in particular is withholding a lot of additional information. They boisterously argue for disclosure, expecting an announcement such as have been carried out by France and the United Kingdom. More on that will be covered later. My guess is that for the most part they would be surprised to learn that the vast majority of the information is already in the public domain.

Summary

There is a huge disconnect between public opinion about the government's role regarding UFOs and institutional responsibilities. The Defense Department engages in matters of national security and no threat has yet emerged from alien invaders. While most of the public believes in UFOs they do not constitute a voting issue. For them they are more of a curiosity than one of intense emotional concern. From a political perspective, being associated with UFOs brings strongly negative reactions from both the media and constituents. To establish an appropriate priority of UFOs, you must ask, "Compared to what?" What will not be funded in order to conduct UFO research? Jobs, health care, and wars rank far higher in the minds of most Americans.

There are many government employees who have seen UFOs. The important issue is to discriminate between their personal observations and beliefs versus the official responsibilities and positions of the agencies they work for. Then there is the long-term secrecy problem. For UFOs to remain under wraps for so long defies the logic of politicians in general, and Washington in particular.

Finally there is the capabilities paradox that must be explained. How is it that the government can handle this unique subject with such precision, yet bungle so many other complex issues? It is time for common sense to prevail.

Since the President of the United States sits at the apex of the government, it is worth discussing what past Presidents have known about UFOs. The next chapter explores what they have and have not said about the topic.

6

THE PRESIDENTIAL PARADOX

I can assure you the flying saucers, given that they exist,
are not constructed by any power on earth.
—PRESIDENT HARRY S. TRUMAN

As the proverbial "Most Powerful Person in the World," a sitting U.S. President should have access to the best information available so that he or she can make informed decisions about the security of the country. Therefore, if the topic is germane, it is of vital importance to understand what information they have concerning UFOs. Knowledge of any potential extraterrestrial intervention would be paramount for anyone in that position.

Several U.S. presidents have expressed interest in UFOs. As Governor of Georgia, Jimmy Carter formally reported seeing a UFO. He later asked about the topic once he was in the White House. Bill Clinton asked Walt Hubbell to look into the subject for him. Governor Ronald Reagan had a sighting from his plane and later commented on the possibility of alien invaders. For these and other presidents, as the Commander in Chief, wouldn't they automatically have access to government knowledge about UFOs?

That depends on to whom you listen. According to the conspiracy theorists, some mysterious *THEY*, an independent, extragovernmental body, makes the determination about which presidents will have access to the knowledge about extraterrestrials, and which will not. Of course who *THEY* are is never articulated but the conspiracy theorists are absolutely sure *THEY* exist. However obtuse such logic may be, it is an essential ingredient in understanding the role of the President of the United States (POTUS) in the whole UFO milieu.

President Jimmy Carter

The most telling presidential story is really that of Jimmy Carter, for it was he who submitted the eyewitness form to the International UFO Bureau. The case was rather mundane, and had it not been that he listed his occupation as "governor," would probably have gone unnoticed in the annals of UFOlogy.

In fact, the sighting actually happened in 1969, two years before he had taken office. The sighting was a simple observation of a light hovering in the sky to the west of his location. Carter stated that he and a group of businessmen were at a Lions Club meeting in Leary, Georgia, when, at about 7:15 P.M., they spotted an unusual object about 30 degrees above the horizon. It was described as a self-luminescent object that appeared to come closer to them, and then stopped above some pine trees. Though Carter slightly varied the details during subsequent interviews, it is consistently reported that there was a bright light that changed colors, including blue, white, and red as they watched. The observation lasted more than ten minutes before they lost sight of the UFO altogether.

The exact date of the event has been debated. While Carter reported it as October 1969, the records show the meeting took place several months earlier. That is not significant except that the skeptics believe that what they actually witnessed was the planet Venus. To confirm or deny that hypothesis, the timing and position of astronomical bodies would be critical. What is more important is that the event made a significant impression on Carter, one that he carried forward for years and would comment on periodically.

Since Carter's observation was on record, UFO enthusiasts were elated when he won the presidency and felt sure that he would ensure that full disclosure of UFO material would be made. As we know, that didn't happen. The story behind what happened related to UFOs while President Carter was in the White House is more interesting than his sighting. According to the official records, his executive office asked NASA to get involved. Uncharacteristically for any agency of the Executive Branch, they did not accept the request. That is nearly unheard of, but points to critical problems when preestablished scientific boundaries are questioned. Obviously NASA pointed to the Condon Report for justification.

The White House Asked for UFO Information

The White House did get officially involved in the UFO topic and the written interactions were often coming from the top of the scientific executives in the government. President Carter's personal role in the inquiry is unclear. However, with public UFO inquiries becoming more frequent at the White House level, it is evident that the White House staff wanted something done to move responsibility to another agency.

The exchange began with a letter from the Executive Office of the President, Office of Science and Technology Policy (OSTP), dated July 21, 1977. The salutations connote the relatively close personal relationship between the people involved. The director of OSTP, Dr. Frank Press, addresses the letter to

"Bob," referring to Dr. Robert Frosch, the administrator of NASA. It is signed simply "Frank." Both were eminent scientists and well known to each other.

In the first letter Press notes that the White House had become a focal point for UFO inquiries and that his office was not equipped to handle the influx. He then stated, "It seems to me that the focal point for the UFO question ought to be NASA." Press recommended a panel be formed to revisit Condon and determine whether or not anything new had come to light. Carl Sagan was listed as a person who might serve on the proposed panel. Press's second suggestion was "that NASA become the focal point for general correspondence and that those inquiries which come to the White House be sent to the designated desk at NASA."

Reviewing these documents, it is interesting to note that neither the Department of Defense nor the U.S. Air Force is even mentioned in the OSTP letter. In addition, there are no markings that indicate that copies were sent to those agencies, or that the letter had been coordinated with them. Such coordination is standard practice when consideration is given to moving responsibility for a subject from one agency to another.

About three months later Dr. Frosch sent Dr. Press a response that was artfully diplomatic and quintessentially bureaucratic. He indicated that "While we are inclined to agree with your recommendations" regarding having a UFO study and focal point, they would need to resolve some questions first. This is a polite way of saying, "No way in hell." He then set forth the issues needing to be resolved, then allowed that if all of the hoops could be jumped through, they would appoint the desired staff officer to handle UFOs. Those hoops included a study concluding that the transfer was justified, and noted there would be a requirement "for additional resources." In English that means "send more money because I'm not paying for this."

In his response to OSTP, Frosch mentioned that NASA already received a few inquiries each month, normally about ten to twelve, and had a form letter that their Public Affairs Office sent out that included an information sheet. Of course the information contained was a summary of the Condon Report. As mentioned earlier, it is hard to estimate the damage that report has done. The sheet ended with a referral to a civilian organization, the Center for UFO Studies in Northfield, Illinois. Again, an action passed is an action handled.

The bureaucratic tango continued when Press sent a letter back to Frosch dated September 14, 1977, stating he was glad they agreed and that he would inform the White House media liaison to send all UFO inquiries to NASA. At this point Press appears to have successfully transferred to mission. But not so fast; that was not at all what NASA had intended. Again while verbally appearing to be compliant, in December Dr. Frosch sent another letter to Dr. Press. This one stated they had thought about it, but barring any new hard evidence they did not believe a UFO study was in the best interest of the nation. He stated that at NASA, "we stand ready to respond to *any bona fide physical evidence from credible sources*" (emphasis in original document). This translated to "you missed this hoop and indicated that NASA was not about to pick up this ball. It was a classic move made by bureaucrats when they say one thing, but have no intention of keeping their word. Studying things to death is a traditional bureaucratic method of appearing to agree, yet then intentionally stalling with intent to kill the action.

There Was a Knowledgeable Insider—Who Quickly Stepped Aside

In the middle of this discussion was interjected Dr. Richard Henry, who had been hired by NASA as the Deputy Director of the Astrophysics Programs. Dr. Henry was relatively new to NASA but also had a long-standing interest in UFOs. For eight years he was a consultant to the Aerial Phenomena Research Organization (APRO). He had also been part of Dr. Allen Hynek's "invisible college," which was a group of qualified scientists who thought that UFOs deserved to be examined properly. In a memo dated October 21, 1977, he advised his superiors of his personal opinions, a few of which included:

> — The UFO-report phenomena exists [*sic*], is widespread, and is of great interest to a large segment of the American people.
> — I see no a priori reason why some of the UFO reports could not be due to sightings of visitors from other worlds or other dimensions.
> — I feel that the Condon investigation did not adequately deal with the UFO phenomenon, and that further government investigation is warranted.

The last paragraph of that memo is quite telling about the mind-set regarding NASA involvement with UFO studies, as well as that of him personally. The archskeptic Phil Klass had already contacted Henry and asked him if he was the NASA point of contact for UFOs. Klass had also provided an admonition about such involvement. In paragraph six Henry states: "*UFOs are (as Phil Klass indicates in a note to me in a copy of his book, which he kindly sent to me) a 'tar baby.' A scientist who touches a tar baby once, as I have, runs the risk of getting deeper and deeper in the goo. I don't have the stomach for it and would prefer to avoid it. But, I also want to make sure that NASA itself does not get badly tarred."*

At this point in the staffing action, we find the official who is probably the most supportive of the UFO phenomena, advising his superiors to avoid the topic. Further, he is in a self-protection mode and states that he wants no official involvement. This is again the negative effect of Condon and an indication of how the scientific community will admonish researchers who dare to venture into the field.

So most amazingly, in the end we have NASA effectively telling the office of the President they weren't going to respond to a White House request. However, in a memo that actually went to President Carter, Henry asked if the President had personally directed the OSTP action to NASA. He indicated there was only one word written by Carter on the side of the memo beside his question—No!

A Personal Request for an Audience

Years later I had a personal follow-up on this issue with President Carter. Dr. C. B. "Scott" Jones had been a friend for several years. In fact, we had even been engaged in an interspecies communications experiment in the Bahamas, along with renowned nuclear engineer Ted Rockwell. Scott was working for Senator Claiborne Pell who had hired him to search out answers to his many questions regarding various phenomena. I had met Senator Pell on several occasions and had discussed many of those issues with him personally. As a result of the Advanced Theoretical Physics project, I had a chance to brief him about UFOs.

Having served five terms in office at that time, Senator Pell was one of the ranking Democrats in the party, one whose requests could not be ignored. Senator Pell sent a letter to ex-President Carter and asked him to receive Scott and me to discuss the UFO topic as we were prepared to fly to Georgia to meet with him. The response was deafening silence. After waiting several weeks, Scott called Carter's administrative assistant and asked if they had received the letter. The answer was yes they had the letter, and "There will be no response." Given all of the commentary that had taken place regarding Carter and UFOs, they obviously did not want anything that would leave a paper trail. Given Senator Pell's position in the Democratic Party, that was quite a snub.

President Dwight D. Eisenhower

Possibly the next most important president in UFO lore is General of the Armies Dwight David Eisenhower. Following his exploits in World War II, he had become known to the American public as simply "Ike." According to the UFO conspiracy legends, Ike met with aliens on one or more occasions. In the most nefarious tales, he agreed to allow aliens to abduct humans for medical experiments. Being a good humanitarian, Ike did require that the humans be returned to their abodes unharmed after the aliens were done with them.

The most frequently repeated story revolves around an incident that took place on February 20, 1954, while the President was vacationing near Palm Springs, California. This story has President Eisenhower being whisked away from the Smoke Tree Ranch where he was staying, and taken to Edwards Air Force Base. Details of the purpose of this trip vary a bit. Some indicate that he was there to view alien bodies and wreckage from a UFO that had crashed. More fanciful descriptions state there were live aliens with "white hair, blue eyes, and colorless lips." These conspiracy theories state that the "Nordic" aliens offered to provide humans their advanced technology provided we would give up our nuclear weapons. Obviously a warrior, Ike refused. According to folklore, it was with the Greys that later he acquiesced to allow medical experimentation. (Of course that is ridiculous.)

It seems true that Ike left the resort that night. The official version states that Ike had chipped a tooth on a chicken bone and went to a local dentist for emergency repairs. He was seen at church in Los Angeles the next morning. For those not familiar with the area, Palm Springs is located only about one hundred miles from Edwards AFB, which is where the meeting would have taken place. Edwards is a very large base and many very sensitive experiments do take place there. It has long runways and huge dry lake beds that are quite hard. Today Edwards Air Force Base is the alternate landing site for the space shuttle.

The reader can determine which version of Ike's itinerary they choose to believe. I find the probability of the alien encounter to be vanishingly small. But for the record, the dentist's name was Dr. Francis A. Purcell, who died in 1974. Additionally, the tooth mentioned in the report was problematic for Ike both before and after this incident.

However, there is a second story about Ike and aliens. This one supposedly took place on February 11, 1955. The location was Holloman Air Force Base in southern New Mexico, just to the east of White Sands Missile Range. This time the actual meeting took place on board the UFO. According to those selling this story, Air Force One landed and was directed to an area well away from the main tower and hangars. Two UFOs were allowed to fly over and one landed close to Ike's aircraft. The second reportedly hovered over the meeting site. A ramp was lowered from the UFO on the ground, and the authors report that the President was seen, albeit by other observers, to enter the craft alone. There is no comment about what the Secret Service was doing during that period. After about 45 minutes, Ike walked out of the UFO and back to Air Force One.

As with the Edwards Air Force Base story, the landing at Holloman has another explanation. The records show President Eisenhower at Thomasville, Georgia, on a hunting trip during the period in question. Conspiracy theorists note that he was not seen by reporters for a thirty-six-hour period and believe that is when he slipped off to Holloman AFB.

One of the people putting forth the information about President Eisenhower's meetings with extraterrestrials is Dr. Michael Salla, a former American University professor who now writes extensively in the UFO field. Salla did note that he wrote to Ike's son, John S. D. Eisenhower, and asked if there had ever been an encounter between his father and ET aliens. The response he reports was brief, "No."

The Eisenhower Connection Gets Even Stranger

If you think the Eisenhower connection could not get weirder, you'd be wrong. In February 2010, one of Ike's great-granddaughters posted material that takes these interactions to a whole new level of improbability. Laura Magdalene Eisenhower posted a long document on the Internet describing her contacts with otherworldly beings. In that posting she stated, "It seemed clear to me that there were both false abductions (more holographic and communicated to the person through chips), and then there were the real ones—the ones I believe my great-grandfather President Eisenhower confronted, which has been a well-known cover-up)." In a tortuous tale that intertwines many myths, she gives further credence to strange events. She goes on to state that her ex-husband was involved and she "came to discover, was giving his sperm to an alien race and had been abducted numerous times." Yes, this is Ike's progeny contributing to the conspiracy theories about him. She also indicated a principal participant in a Mars mission was none other than Dr. Hal Puthoff, a friend who was mentioned previously regarding our meeting with Dr. Edward Teller.

Generally overlooked is what Eisenhower said when he was asked a direct question about UFOs in December 1954. Ike answered, "Well, with regard to those recent reports, nothing has come to me at all, either verbally or in written form. And I must say, when I go back far enough, the last time I heard this talked to me, a man whom I trust from the Air Forces [*sic*] said that it was, as far as he knew, completely inaccurate to believe that they came from any outside planet or otherwise." If President Eisenhower is to be believed, this suggests that by 1954, UFOs were not receiving the attention necessary to bring the topic to his level.

President Bill Clinton

President Clinton privately expressed some interest in UFOs, but certainly had no special knowledge. In his memoirs, a close confidant of the Clinton's, Webster "Webb" Hubbell, wrote that while working in the White House Clinton had asked him to look into two of the most controversial conspiracies of the United States. One was the enigmatic tangle that will likely be debated for decades, who really did kill Kennedy? The second was to find out what the truth was about UFOs. According to Hubbell, he failed on both counts, and was unsatisfied with the responses he received. Of course, by that time the Clinton administration was embroiled in a number of scandals and there was little time for trivia.

There is a story about a boy in Belfast, Ireland, named Ryan asking Clinton about the Roswell incident. Clinton allegedly responded, "No, as far as I know, an alien spacecraft did not crash in Roswell, New Mexico, in 1947." He went on to say, "And, Ryan, if the U.S. Air Force did recover alien bodies, they didn't tell me about it either, and I want to know." At least these stories are consistent, and probably true. Of course the conspiracy theorists will claim it is an example of the information being withheld intentionally.

Laurance Rockefeller Raised the Issue

During the Clinton Administration there was another assault on the White House to obtain information about what was known officially regarding UFOs. The entire sequence of events is well documented on Steve Bassett's Web site, *www.paradigmresearchgroup.org*. As you may remember, Bassett was the lone UFO registered lobbyist. At the center of the action was Laurance Rockefeller, a wealthy philanthropist with a penetrating interest in UFOs who was willing to put his money where his interests lay. It was Laurance Rockefeller who later funded the conference for scientists at the Pocantico Conference Center near Tarrytown, New York, that was organized by Dr. Peter Sturrock.

As a major contributor to the Democratic Party, Laurance Rockefeller could not be ignored by the Clinton staff. As money does buy access, Rockefeller was also able to talk directly to the Clintons about the topic. As with the Carter administration, the Office of Science and Technology Policy (OSTP) became one of the main focal points. Dr. John Gibbons, whose title was Assistant to the President for Science and Technology, was a logical choice as a point of contact for Rockefeller and the people who helped support his effort. While attempting to gain the interest of the staff, Rockefeller financed a book called *UFO Briefing Document: The Best Available Evidence*. It was coordinated by Marie "Bootsie" Galbraith and Antonio Hunees and written by Don Berliner and was an excellent compendium of UFO cases. The book received good reviews by the UFO community, but did little to sway the scientific establishment.

Since OSTP is relatively small, Gibbons reached out to other agencies to learn what they knew about the topic. Among those responding was the then Secretary of the Air Force, Dr. Sheila Widnall. In her letter to Gibbons on May 20, 1994, she wrote, "My policy is that we are to declassify everything even remotely related, and anything our people think still needs to be classified will have to be justified to me." Those familiar with the formal U.S. document that was released a short time later (in 1997) and titled *Roswell Report: Case Closed* will note the information to Dr. Widnall came from the same Colonel Weaver's office that was doing that rather dreadful report. The memo dated May 20, 1994, does elucidate a few issues worth noting. Of course it was negative in tone and stated, "The review to date has not indicated anything out of the ordinary happened at Roswell. We can rule out an airplane crash: probably rule out an errant missile test." The memo goes on to address the Project Mogul balloon theory that they banked on in the final report. The memo also notes the requirement for Weaver's staff to get additional clearances. That probably refers to the Department of Energy and the Q clearance system they use that also establishes Sigma levels for various aspects of nuclear weapon design. This is indicative of the problems that exist between DoE and DoD in that they will not accept the clearances from the other agency on all issues.

Melvin Laird, the former Secretary of Defense under President Nixon, was also contacted by Laurance Rockefeller. His response is quite interesting as he is one of the few senior officials who seems willing to admit any personal knowledge about UFOs. In part, Laird's letter says, "I am sure that should classification be lifted, some individuals will be disappointed as certain of these phenomena are pretty well explained. Any review will disappoint some individuals who have built up some rather extreme antidotal [*sic*] and uncollaborated [*sic*] accounts which the removal of classification might discredit to a large extent. Removal of undue classification will remove the speculation of some of these reports." I believe Laird meant to say anecdotal and uncorroborated reports.

John Podesta Goes Public

During the Clinton administration John Podesta held a number of key positions, including Deputy Chief of Staff, Senior Policy Advisor on Government

Information, and finally Chief of Staff of the White House. While serving in the White House he was a proponent for the declassification of many of the millions of documents that were then held. Since leaving those positions Podesta has made several statements supporting the release of classified UFO documents. At a conference held by the Coalition for Freedom of Information in 2002 he said, "It is time for the government to declassify records that are more than 25 years old and to provide scientists with data that will assist in determining the real nature of this phenomenon."

A few years later, at another conference on September 29, 2007, he stated, *"I think it's time to open the books on questions that have remained in the dark; on the question of government investigations of UFOs. It's time to find out what the truth really is that's out there. We ought to do it because it's right; we ought to do it because the American people quite frankly can handle the truth; and we ought to do it because it's the law."*

What is not clear from these statements is what, if any, attempts he made while at the seat of power to gain access to this information. The reality is that the information flow is phenomenal and keeping track of critical items is at best difficult. There is not much time for exploration of personal interests.

From the information available, it is clear that during the Clinton administration there was interest by both government officials and UFO enthusiasts to acquire information about this topic. It seems equally clear that despite a fairly concerted effort, nothing of significance came to light. Conspiracy theorists will suggest that this proves that *THEY* have total control of the situation and can determine which presidents may have access. A more rational consideration might be that Melvin Laird was right and there really isn't much information available, and the classified information that does exist would be disappointing to the public.

In a foreword to Leslie Kean's 2010 book *UFOs: Generals, Pilots, and Government Officials Go On the Record,* Podesta supported a recommendation for the creation of a new U.S. Government agency to study UFOs and interact with foreign governments on the subject. Coming from one as politically astute as he, that is a very strange endorsement. He surely knows that given the current economic environment, with substantial cuts to existing agencies, snowballs have a better chance in hell.

President Ronald Reagan

It has been written that Ronald Reagan had at least two UFO sightings. There was an early one that Lucille Ball told about in Jim Brochu's *Lucy in the Afternoon: An Intimate Memoir of Lucille Ball.*

The second sighting occurred in 1974 while Reagan was the Governor of California. This story has been backed up by Colonel Bill Paynter, who had retired from the U.S. Air Force and was Reagan's pilot while he held that office. The reports have four people aboard the plane: pilot Bill Paynter, two security guards, and the Governor of California, Ronald Reagan. While the quotes are attributed to the *National Enquirer* they have been confirmed by multiple sources as various UFO investigators have chased this story. There are several accounts published about Reagan describing the event.

Paynter recounted the details as follows:

> We were flying a Cessna Citation. It was maybe nine or ten o'clock at night. We were near Bakersfield when Governor Reagan and the others called my attention to a big light flying a bit behind the plane. It appeared to be several hundred yards away. It was a fairly steady light until it began to accelerate then it appeared to elongate. The light took off. It went up at a 45-degree angle at a high rate of speed. Everyone on the plane was surprised. Governor Reagan expressed amazement. I told the others I didn't know what it was. The UFO went from a normal cruise speed to a fantastic speed instantly. If you give an airplane power it will accelerate but not like a hot rod and that is what this was like. We didn't file a report on the object because for a long time they considered you a nut if you saw a UFO.

Paynter added the UFO incident didn't stop there. He stated that he and Reagan had discussed their UFO sighting from time to time in the years following the incident.

It was indicated that about a week later Reagan mentioned the incident to a reporter, Norman C. Miller, then Washington bureau chief for *The Wall Street Journal*. Reagan told Miller, "We followed it for several minutes. It was a bright white light. We followed it to Bakersfield, and all of a sudden to our utter amazement it went straight up into the heavens." This story has Reagan seemingly not aware that he was talking to a reporter and would likely be taken as on the record. We do know that both Ronald and Nancy Reagan were open to studies of phenomena and the press would excoriate them for it once they found out.

Did Belief in UFOs Impact Policy?

Whether or not his personal experiences with UFOs influenced his policy statements is hard to say. However, the notion of an alien threat was a recurring theme in his presentations. In a famous speech before the United Nations in 1987, President Reagan stated," "When you stop to think that we're all God's children, wherever we may live in the world, I couldn't help but say to him, just think how easy his task and mine might be in these meetings that we held if suddenly there was a threat to this world from some other species from another planet outside in the universe. We'd forget all the little local differences that we have between our countries and we would find out once and for all that we really are all human beings here on this earth together."

Some conspiracy theorists hypothesize that President Reagan had knowledge of such a threat and was warning the world about it. They also note that it was under his administration that the Strategic Defense Initiative (SDI) or Star Wars was born. Some of them suggest that the real purpose of Star Wars was to fight extraterrestrials when they attempted to invade Earth.

Earlier the Star Wars program was addressed. It had nothing to do with an ET threat, nor was that even given consideration in the program. It is more likely that it was a strategic initiative, and one that broke the back of the USSR as they could not keep pace with the expense of that program. Further, if you actually watch the recording of that UN speech, there is not a hint of threat in Reagan's voice. Rather, he seems to be imploring the people of the Earth to come together and stop the consistent warfare in which we so frequently engage.

President Gerald Ford

Historians are aware that President Ford was the only person to attain that position without being elected and served only part of one term following Nixon's departure. While he is not known to have addressed the UFO issue while President, as a Michigan Congressman he did formally make a request that information be released. In a letter dated March 28, 1966, to Mendel Rivers who was then the Chairman of the Armed Services Committee, Ford wrote,

> No doubt, you have noted the recent flurry of newspaper stories about unidentified flying objects. I have taken special interest in these accounts because many of the latest reported sightings have been in my home state of Michigan. . . . Because I think there may be substance to some of these reports and because I believe the American people are entitled to a more thorough explanation than has been given them by the Air Force to date, I am proposing that either the Science and Astronautics Committee or the Armed Services Committee of the House schedule hearings on the subject of UFOs and

invite testimony from both the executive branch of the Government and some of the persons who claim to have seen UFOs. . . . In the firm belief that the American public deserves a better explanation than that thus far given by the Air Force, I strongly recommend that there be a committee investigation of the UFO phenomena. I think we owe it to the people to establish credibility regarding UFOs and to produce the greatest possible enlightenment on this subject.

Congressman Ford was successful in obtaining the hearings he requested. On April 5, 1966, U.S. Air Force representatives, including Allen Hynek, did appear before Congress and testified. Two years later, on July 29, 1968, another session was held. There were six scientists making presentations, and several were insistent that the UFO issue deserved serious consideration. There was one who, despite acknowledging that there were cases that were unexplained, was quite skeptical—Carl Sagan. Shortly after that hearing the Condon Report was released. In the end, very little came of this effort, and there are no indications that Ford followed up with inquiries when he became POTUS. That lack of inquiry probably speaks volumes about the relative importance of UFOs and the national stage.

Confronting *THEY*

What has been articulated here, and is found in greater depth in research projects devoted to the study of the presidency and UFOs, is a clear indication that the topic has been raised on several occasions by those in authority. Yet there is no indication that any of them have been successful in finding any significant information about the topic. This means that the POTUS, just like everyone else in the American public, has heard rumors about crashes and secret programs, but when they drill down, there is little evidence to support the stories. Secrecy is indeed a problem. When there is a lack of transparency, people will fill in the voids, largely based on their imagination.

There are those who state that UFOs are an *unacknowledged special access program (USAP)*. That means that even the existence of the program is denied, let alone the facts associated with the project. In vacuous statements, these conspiracy theorists claim that members of the Executive Branch of the U.S. Government who work on these programs are required to lie to everyone, including the POTUS, about the existence of that program—in this case UFO research. This is both utter nonsense and illegal. For conformation, just read the oaths that are sworn or affirmed when assuming positions of authority and responsibility.

The requirement comes from Title 5, Section 3331 of the U.S. Constitution and states: "An individual, except the President, elected or appointed to an office of honor or profit in the civil service or uniformed services, shall take the following oath: 'I, . . . do solemnly swear (or affirm) that I will support and defend the Constitution of the United States against all enemies, foreign and domestic; that I will bear true faith and allegiance to the same; that I take this obligation freely, without any mental reservation or purpose of evasion; and that I will well and faithfully discharge the duties of the office on which I am about to enter. So help me God.' This section does not affect other oaths required by law." Note the oath is absolute and that there are no provisions or exceptions interjected by *THEY* who allegedly hold the keys to some USAP.

Quintessential October Surprise

There is a significant logic flaw in the notion that some Presidents have had access to the information that ET exists, while others do not. That flaw is in the election process. As noted earlier, an official's most important duty is to get elected. After that it is to be reelected. There have been many close presidential elections. The younger generations believe that John F. Kennedy was swept into office. In reality in the 1960 election he narrowly defeated Nixon with a plurality of 49.7 percent to 49.6 percent of the popular votes. In the 2000 contest between George W. Bush and Al Gore, the winner acquired only 47.9 percent of the popular vote, while the loser held 48.4 percent. Of course the difference was in the electoral vote, in which the Supreme Court held Florida for Bush, giving him a 271 to 266 vote victory.

The point is that every four years the United States holds a presidential election to determine who will ascend to the most powerful position on the planet. It is inconceivable that a party or organization having knowledge of an extraterrestrial civilization or threat would not play that card to ensure victory in a tight election. *October surprise* is a term used to express concern that a candidate will either create, or hold back, an event that will precipitate a response in late October of an election year. The intent of that event would be to swing the election and not allow the opposition to have time to respond.

The 1980 election is sometimes pointed to as an example, albeit one that didn't work. Reagan feared that Carter might engineer a release of the Americans being held hostage in Iran to boost his sagging approval rating. As it turned out, the Iranians, angry with Carter for the Desert One rescue attempt, stated they would not release the hostages as long as Carter remained in office.

When fear is invoked, the incumbents gain. It is the "don't change commanders in a war" ploy. Additionally, the percentage of change can be quite small and still be effective. A mere one or two point change could have a dramatic effect on the outcome of an election. Some naïve conspiracy theorists claim that knowledge of UFOs or proof of ETs is so special that it is immune to human frailties and would not be used by either side. To say that this is implausible is an understatement. Just note the dirty tricks and vicious lies that have been used to attain office and it becomes clear no topic is off limits.

Considering the potential for an October surprise before the 2008 election, an e-mail was posted in response to claims that *THEY*, including elements of the U.S. Government, had successfully reverse-engineered a UFO. While not true, the topic is germane to the POTUS discussion, for if *THEY* do exist, that must include heads of industry and/or an extremely wealthy subset of the population. Change would not be welcome, and that was the primary theme of Barack Obama's campaign. Therefore, it was stated that if *THEY* did exist, this was the time that *THEY* would move. That a UFO disclosure did not happen was really a challenge to the conspiracy theorists.

The following is the text of an e-mail that was published in late October 2008. It details a challenge to the UFO community about why, if *THEY* held the secret, this would have been the time that the information was released. They did not respond. Here is the content of the e-mail:

FTR: It seems that the truth about ARV should be revealed by Tuesday, November 4, 2008. It is the quintessential October (now November) surprise. Facts are as follows:

— United States (and the rest of the world) is in deep economic trouble
— The polls show the race tightening, but McCain generally still behind
— CEOs of major U.S. companies heavily favor McCain
— Energy issues are significant and all parties acknowledge the need for new sources
— A change in energy posture would have extremely positive impact on the US & global economy
— Republicans have favored the big energy companies
— The CEOs (*THEY*) would believe there is an urgent need to ensure the election goes in their favor (i.e. McCain)
— Incumbents are favored in time of crisis

— McCain is viewed as more experienced in military matters
— Revelation of ET/ARV would have an inherent threat component
— Revelation of ET/ARV would be sufficiently significant to ensure the election of McCain

Therefore:

— If ARV exists, it will be revealed this weekend (or by Monday)
— If it is not revealed by Tuesday, it is reasonable to assume it (UFO knowledge) does not exist within USG (or other organizations)

Just thoughts,
John Alexander

The POTUS Briefing

"Be careful what you wish for" is an old adage that applies to UFO enthusiasts who enter into the Washington political arena as naïve lambs to the slaughter. We have already documented how UFO supporters have approached the White House in the past. The results have been minimal. Despite previous failures, it is happening again. The credulously led people have little concept of how the government functions and the powers that are invested in the President. They seem to believe that any POTUS can wave a wand and have old information revealed and new programs started. There appears to be considerable confusion between the authority of the POTUS and that of a potentate. The topic can be found on the Disclosure Project Web site, but they are not alone. There are other groups developing a POTUS brief as well. In fairness, this was essentially the same approach that was taken by Laurance Rockefeller with the *Best Evidence* book that was designed to achieve the same results.

It is important to understand that many of the people in these groups believe that the U.S. Government, or *THEY*, has reverse-engineered a UFO. Remember the comments of the Honorable Paul Hellyer, former Minister of Defense of Canada, who confirmed those claims, although with scant evidence. There are probably millions of Americans who also believe this notion.

Why then would an admonition concerning wishes for a high-level briefing be issued? It is so that one can rationally consider the consequences, both positive and negative, of raising the UFO topic at the highest levels of government, especially without doing the considerable groundwork beforehand.

Proponents will argue that adequate preparation has been done. However, these efforts are grossly inadequate for the agenda they are pushing.

When President Obama assumed office in January 2008 his approval rating was soaring and a large majority of the public, both here and abroad, looked forward to dramatic change in governance. Included were many UFO aficionados who thought he might be the President that would fulfill their wishes and release the store of information they believed to be held by the government. To the disappointment of many citizens, government transparency in the new administration barely improved regarding any topic.

However, in the past two years Obama's political fortunes have changed markedly as he has seen his approval ratings plummet while coupled with rapidly increasing disapproval ratings. Of all the actions he might take in order to gain favor with the American people, officially acknowledging UFOs is not among them. Certainly he and his advisors know that supporting the creation of the proposed new agency to study the topic would be absolute career suicide.

UFOs and the Manhattan Project: A Comparison

For the moment let's set aside the overwhelming factor of competition for attention with every other strategic problem that demands attention—little things like health care, the economy, wars, not to mention maintenance of credibility for reelection. For comparison of how the project might progress, the decision to build the atomic bomb in the Manhattan Project during World War II can be drawn upon. There are some useful parallels to consider. First, the project was supported by many in the scientific community, but not by all. The champion selected to deliver the concept to President Roosevelt was none other than Albert Einstein, then revered as one of the greatest scientists who ever lived. However, even he did not have access to the President and the task fell to Alexander Sachs, an unofficial advisor to Roosevelt. Einstein had assistance in drafting the letter. It was really Leo Szilard who had become most concerned about the German efforts to build an atomic device. Szilard secured assistance from Edward Teller and Eugene Wigner, also scientific luminaries, in preparing the letter and on August 2, 1939, Einstein signed two versions that were delivered to Sachs.

One month later, on September 1, 1939, Germany invaded Poland and World War II officially began, placing increased emphasis on international security matters. On October 11 Sachs met with Roosevelt and presented Einstein's letter. While most people now believe that this was a turning point in the development of nuclear weapons, that is far from the truth. Roosevelt opted to have studies conducted about the feasibility of developing such a weapon. At the time most American physicists did not believe that the German concept posed a threat, or that the theory was viable. In fact, little happened for two years and only $6,000 was allocated to buy graphite and uranium for experimental purposes. Then, on December 6, 1941, the U.S. Government decided to begin the atomic bomb but it did not evolve into the Manhattan Project until August 1942.

There is no doubt that the development of nuclear weapons was one of the most significant decisions ever made. It was conceived of and brought forward by some of the brightest minds America had to offer; all were well established and highly respected. Their concept was still not rapidly accepted, and the bureaucratic response was predictable—study it some more! In this most urgent case POTUS did not just direct a project to begin and provide funding for it.

Considering this example, the question to those who propose briefing POTUS on UFOs is, which Einstein do you have that will carry the concept forward? Next, are you prepared for the consequences? No one just briefs POTUS on new concepts until they have been thoroughly vetted. That means whoever does carry the ball will be forced to conduct prebriefings, probably many of them, before an audience is even granted. Those being briefed will have agendas quite different from the proponent's. Technical issues aside, there will be fear of political fallout from the very notion that such a briefing ever reached the White House, let alone POTUS. It is highly likely that the skeptic's organizations, such as the Committee for Skeptical Inquiry (CSI), will learn of the attempt and launch a counteroffensive. Unbounded by facts, they aggressively tackle any perceived threat with a media campaign, making political viability even more treacherous for the White House.

At any point in this process, the easiest and safest response by staffers is a simple, even polite no. If sufficient political strength exists, such as a Laurance Rockefeller, then the next step would be to recommend an independent study of the subject. While that may sound reasonable it is most likely that the organization chosen for such a task would be the National Academy of Sciences. While they have many excellent scientific experts on call, this is one of the most conservative scientific bodies in the world. The participants did not get to their positions by supporting unproven alternative theories or phenomenology. Most likely none of them will have ever read the Condon Report, but they will have drawn conclusions based on what they have heard. In addition, to be independent in studies of phenomenology really translates to having no prior knowledge of the topic.

Prior Experience

This postulated scenario is not conjecture on my part, but based on personal experience. In 1984, the U.S. Army asked for a study on the human potential techniques that were being employed by various units. In 1988 the National Research Council (NRC) published their report entitled, "Enhancing Human Performance," which proved to be highly negative about almost all subjects discussed. I addressed the panel on several occasions. Frankly, there was an appalling lack of fundamental knowledge about the subjects under review. Only one person on the panel had ever studied, or written about, any of the topics. That was psychologist Ray Hyman who was a founder of the Committee for Investigations of Claims of the Paranormal, the forerunner of CSI. A personal request by Senator Claiborne Pell was ignored and the National Research Council refused to appoint even one fully qualified proponent member to the panel. As the proverbial one-eyed king, Hyman became a virtual leader of the study and the person the group turned to for explanations.

Should the NRC, or any similar organization, be contracted to perform a study on UFO phenomena, that is exactly what will be done again. Qualified proponents would be excluded because they are perceived as biased, while similarly credentialed opponents would be allowed as having critical thinking skills. The bottom line to such an independent study is the most likely outcome would be Condon II.

It is my opinion that the risks for requesting an official independent study of aerial phenomena would be great. Given the amount of nonsense that is published in the area, researchers could legitimately concur with Condon, correctly noting that the vast majority of the reports can be attributed to conventional answers, often just misidentification. The real concern is that the report could easily be construed to denigrate high-risk/high-reward research into alternative energy mechanisms.

What a Flow Chart Would Look Like

This flow chart depicts a very simplified version of how the process to brief POTUS would work. First, the persons having the briefing document must find an influential individual with the connections necessary to make an appointment. This meeting would probably be with a staff person who would determine whether or not the briefing should go forward. If rejected, the process ends. If approved the briefing goes forward to the next level. Note this is probably an iterative process with the possibility of rejection at any juncture.

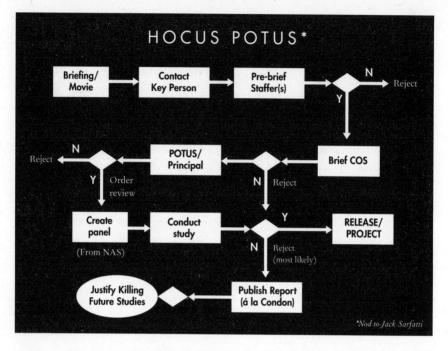

A final prebriefing is likely to be to the Chief of Staff who would personally approve the briefing moving on to POTUS. If lucky, a condensed version is made in person. If not, the official sidestep is to advise you to leave the presentation and a staffer will conduct it later. Of course that is impossible, as they do not know the material well enough to provide the presentation and answer questions with any degree of subject matter expertise.

It is highly unlikely that POTUS would approve anything outright. The best result to hope for is a referral to a study panel. That panel then conducts a study. If a miracle happens the results are positive and project is authorized. As will be explained, the troubles have just begun at that point. The most likely study outcome would be a report à la Condon II and the field is set back at least a decade. Then the findings from the study are then used as justification to kill other alternative proposals.

Should the project be approved, it would then be assigned to some organization to execute. It is unlikely that any organization would be happy to get this assignment. They are already running faster than humanly possible, and this is just another burden. If the project was not resourced, then the organization must take support out of its hide—meaning transferring funds or people from other already approved projects. If the recipients of the project, and intermediate supervisors, have not been involved in the approval process, you have probably created enemies, ones who must now produce the results that are hoped for by the UFO enthusiasts. The B-Team that was bypassed in the scramble for high-level intervention will get revenge.

Summary

Many presidents have expressed interest in UFOs, yet none seem to receive a satisfactory answer regarding their nature or origin when they have made inquiries into various departments within their government. To suggest that the information is intentionally withheld is specious. The most prosaic answer is that those departments don't have the information to provide. Moreover, the scientifically oriented organizations are tainted by the Condon Report, and in general, they simply do not believe there is anything to UFO reports.

The notion that secret subelements of the U.S. Government must deny the President access to such information when requested is both false and illegal. The information available suggests that even presidents have heard the same rumors and innuendoes that have the general public and operate on their preconceived ideas.

It is important to understand that those who believe a POTUS can, or would, wave a wand and make the information about UFOs public are terribly naïve. They simply do not comprehend how large organizations work. No matter the importance of the topic, there are procedures that are followed. A politically sensitive tar baby like UFOs would be treated with great care. There is no doubt that if there was a decision to proceed with UFO research, it would be assigned to the most conventional scientific bodies for execution. For those who would choose to pursue that course of action: be careful what you wish for. The most likely outcome would be a study that sets the field back decades.

According to the conspiracy theorists, the entire UFO field is dominated by a mystical organization known as Majestic 12 and those twelve even decide whether or not POTUS is briefed. Mythically, they appear to be ubiquitous and near omniscient, and certainly worth examining in the next chapter.

7

MAJESTIC 12

We have learned in recent years to translate almost all
of political life in terms of conspiracy.
—JOHN LE CARRÉ

At the beginning of the book it was mentioned that it was important to understand conspiracy theory if you want to study UFOs. The conundrum can be described quite simply. Either you believe that some secret organization is concealing the information or you are seen as part of the cover-up. Of all of the conspiracy theories related to UFOs, none is more pivotal than Majestic 12.

UFO lore is thoroughly infused with the mysterious Majestic 12. According to the legend, MJ-12, as it is better known, a secret cabal comprised of influential scientists and politically connected intelligence heads, forms the oversight committee that controls all of the clandestine UFO material. Postulation about MJ-12 stems from a number of supposedly leaked documents. The origin of the term is not clear, but first use is sometimes attributed to a teletype dated November 17, 1980, and seen by some researchers early the next year. Since then debate over authenticity has raged among UFO researchers with many of the best minds declaring the material to be deliberate counterfeits. Nonetheless, the conspiracy has taken on a life of its own. For the record there is no truth to the matter, but I am regularly accused of being part of this organization. My response has been to ask about the whereabouts of my paycheck that has never appeared and neither am I involved pro bono.

MJ-12: The Beginnings

MJ-12 came into private view in late 1984, albeit under unusual circumstances. A roll of undeveloped 35 mm film came via the U.S. Postal Service to Jamie Shandera's home. The film that arrived in his mailbox inside a plain brown envelope had no return address, but bore an Albuquerque postmark. It was only after the film was developed did Shandera learn the contents. Deposited on the negative were a series of papers that would become known as the Majestic 12 Documents. Shandera was working with Bill Moore, who in 1980 had written *The Roswell Incident* about the famous New Mexico UFO crash. Timothy Good included copies of some of those documents in a book published in the UK. That triggered Moore to publicly announce their existence in the United States in 1987. Eventually Moore made the contents of the documents known to anyone who was interested. These documents are available on the Internet and are covered in fair detail in Stanton Freidman's book *Top Secret/ Majic*, which he published in 1996.

The promulgating document allegedly comes from President Harry Truman in a September 24, 1947, memorandum for Secretary of Defense James Forrestal. It mentions a conversation between Secretary Forrestal and President Truman and directs that "this matter shall be referred to only as Majestic Twelve." It also reserves the right of disposition to the Office of the President, and brings Dr. Vannevar Bush, the Director of Central Intelligence, into the loop.

The key document is known as the *Eisenhower Briefing Document*. Dated November 18, 1952, this eyes-only paper was allegedly prepared for President Elect Dwight Eisenhower. It told him that "Operation Majestic-12 is a Top Secret Research and Development/Intelligence operation responsible directly and only to the President of the United States." In addition, this paper had a list of names of the membership of MJ-12 as it was initially formed. They included:

— Adm. Roscoe H. Hillenkoetter
— Dr. Vannevar Bush
— Secretary James V. Forrestal*
— Gen. Nathan F. Twining
— Gen. Hoyt S. Vandenberg
— Dr. Detlev Bronk
— Dr. Jerome Hunsaker
— Mr. Sidney Souers
— Mr. Gordon Gray
— Dr. Donald Menzel
— Gen. Robert M. Montague
— Dr. Lloyd V. Berkner
— Gen. Walter B. Smith (who was added after the death of
        Forrestal)

---

* Forrestal died in 1949, at which point General Walter B. Smith joined the MJ-12.

Certainly this would have been an august body, one worthy of addressing the most pressing needs of the nation. Those familiar with history will recognize the strong connections to the U.S. Air Force and the Intelligence Community. Souers, Vandenberg, Hillenkoetter, and Smith, respectively, were the first four Directors of Central Intelligence, while Gray, Hunsaker, and Menzel all previously held intelligence advisory positions. Berkner had participated on the CIA Robertson Panel.

MJ-12 Versus Watergate

It would be an understatement to suggest that the process for release of the MJ-12 documents was extremely strange and makes little sense. For purposes of discussion of the elements that make the whole Majestic scheme highly unlikely, the comparison with the Watergate scandal that led to the demise of Richard Nixon as President of the United States is enlightening. The underlying similarity is the notion that someone highly placed in the U.S. Government wanted to get very sensitive information into the public domain. In the case of Watergate, they were quite successful. For MJ-12, except for the UFO community, few people have ever heard of this conspiracy-based organization.

The High-level Source Was Known

The differences are enormous, and worth considering when attempting to evaluate the validity of the Majestic documents. We now know that *Deep Throat*, the code name that *The Washington Post* attached to the original source, was Mark Felt. At the time Felt was the Deputy Director of the FBI. Of course the Deep Throat moniker related to the famous porno movie of that name.

Felt's motive for becoming an informant for the media is plagued with controversy. A longtime FBI executive, Felt had been very loyal to the founding director, J. Edgar Hoover. After Hoover's death, L. Patrick Gray, a lawyer with no experience in the FBI, was appointed as the new director. Those people who suggest that Felt's motive for divulging the information to the *Post* was personal note that he did not approve of an outsider assuming the leadership of the Bureau. Felt and other agents believed that the position should have gone to someone with extensive law enforcement experience, possibly Felt himself. Many other observers believe that Felt's main motivation was patriotic. The Watergate break-in had despoiled the Office of the President, and the Department of Justice was not doing enough to address the problem.

Whatever Felt's motives, he was clearly in a position to have access to the sensitive data he was revealing. Next, Felt decided to provide the information to two young but experienced investigative reporters. He chose Bob Woodward and Carl Bernstein based on their assignment to investigate the Watergate burglary in a major media venue; a perfect fit for this venture. Additionally, Felt had previously been in contact with Bernstein on other routine matters. Therefore, one of the two reporters was known to him and considered reliable. In order to pass information, physical, albeit clandestine, meetings were arranged. The reporters were not to call his office or be seen in public with him. The initiation of those meetings was right out of spy novels. Felt would leave a flowerpot in a designated location, and then he would deposit newspapers with times circled on inside pages. The physical contacts came very late at night, and reportedly often took place in an underground garage in Roslyn, Virginia.

While the public did not know who Deep Throat was, key management personnel in *The Washington Post* did, including Ben Bradlee, then the editor, and Leonard Downie, Jr., his successor. The circle was very tight and according to Woodward only seven people actually knew Felt's identity.

The Washington Post Had Credibility and Resources

Obviously the choice of *The Washington Post* provided a degree of credibility and extensive resources for the conduct of the investigation. It was the most powerful media outlet in the area and had national readership with many of its investigative stories being syndicated worldwide. It also had a long history of tackling difficult and sensitive issues in the capital region and had experience in handling explosive exposés.

Most of the time Felt merely provided clues to Woodward and Bernstein, who would then have to find the people involved and interview them. To support their assertions, the reporters were required to develop independent corroboration, usually from multiple sources. If verification was not available, allegations were not made. The investigation lasted two years and terminated with several officials going to jail, and the first and only resignation of an American president.

But those events did not stop the speculation about the identity of Deep Throat. There were many suspects, several of whom had both access and motive for releasing the information. Finally, in 2005, at age ninety-one, Mark Felt acknowledged that he had been the source of Watergate information for Woodward and Bernstein. It was only after Felt's public statement and a large article appeared in *Vanity Fair* that those who had been in the know on *The Washington Post* confirmed the information.

Now the reader should compare that release of sensitive information with how MJ-12 has proceeded. Begin with the source. As far as is publicly available today, the source of the documents remains unknown to everyone, including the people who released them. Rather than meeting under a bridge late at night, the perpetrator sent an anonymous envelope, not even indicating what the film related to, or even the general topic. There have been no face-to-face meetings between the reporters and their confidential source, nor was there any way to vet him or her.

Rather than approaching *The Washington Post*, *The New York Times*, or some other established and credible media outlet with the information, the mysterious MJ-12 surrogate Deep Throat chose to send it to people with very limited experience in reporting and no support mechanism to get the information to the general public. While the documents begged for critical analysis, neither Moore nor Shandera had resources comparable to *The Washington Post* with which to follow up.

Further, they certainly did not have the means to disseminate their information and get it into mainstream American consciousness. If these sources are to be believed regarding intent, they were not just titillating the already convinced UFO community, but rather the focus was to begin preparing the public for the release of some momentous acknowledgment—that ET was real.

There were other MJ-12 documents released in an equally inglorious fashion. In 1992 documents allegedly were deposited in the mailbox of Tim Cooper at Big Bear Lake in California. The origin of the documents and rationale for this mode of release remains a riddle—one without a good answer.

Then in 1994 photographs of an entire manual purporting to be a guide for Special Operations arrived addressed to known UFO researcher Don Berliner. As with the other documents, there was no return address. However, Dr. Bob Wood was able to track the postmark to Quillin's Drug Store in the town I grew up in—La Crosse, Wisconsin. Again the document was in the form of a roll of undeveloped 35 mm film. The main thesis of the document seemed to be to verify the former documents and to support the Roswell crash story.

The title of the manual was *Extraterrestrial Entities and Technology, Recovery and Disposal*. It's annotated as *Majestic-12 Group Special Operations Manual, SOM 1–01*, and dated April 1954. Most UFO researchers dismiss this document as a total hoax. They include Don Berliner, to whom the package was addressed. Interestingly, both Bob and his son, Ryan Wood, spent considerable time and effort in attempting to authenticate this product and have come to another conclusion. Based on extensive analysis, in which they examined many of the arguments against authenticity, Bob Wood is very sure that the manual was written in 1954.

However, that does not make it a real UFO document. The explanation Bob told me was that he believes the manual was created at the time indicated, but with the intention that it be leaked to Soviet spies as part of a disinformation campaign. As far out as that may sound, it makes more sense than the notion that some special operations unit was training to recover aliens. Both the United States and the Soviets conducted disinformation operations that were designed to mislead the other, frequently with the intent of causing the opponent to commit valuable resources to wild-goose chases.

The CIA and MJ-12

The MJ-12 issue has been raised with the CIA. In an extensive review of the agency's involvement in UFOs, Gerald Haines, the historian for the National Reconnaissance Office, addressed some of these documents with very firm evidence that they were fraudulent. His report, *CIA's Role in the Study of UFOs, 1947–90*, ends with this statement: *"Dr. Larry Bland, editor of The George C. Marshall Papers, discovered that one of the so-called Majestic-12 documents was a complete fraud. It contained the exact same language as a letter from Marshall to Presidential candidate Thomas Dewey regarding the 'Magic' intercepts in 1944. The dates and names had been altered and "Magic" changed to 'Majic.' Moreover, it was a photocopy, not an original. No original MJ-12 documents have ever surfaced."*

As with all of the MJ-12 documents provenance remains a critical and unexplained problem. The best answer would appear to be that there are people with too much time on their hands.

The Aviary

One of the Majestic curiosities was the evolution of a mysterious group known to UFO enthusiasts as the Aviary. According to the lore, members of the Aviary were insiders that had access to the government's knowledge of UFOs. The reality is quite another story, and actually rather laughable.

I do admit to being a member of this group. In fact, I was officially anointed with Hal Puthoff and Scott Jones at a meeting in Dayton, Ohio. While Bob Collins, in his book *Exempt from Disclosure*, called the meeting, "A summit without cocktails," it was really quite mundane. Among the others present were Bill Moore and Jamie Shandera who actually controlled the whole Aviary concept. Since Moore and Shandera frequently communicated by commercial telephones, and were being a bit paranoid (which goes with the territory), they wanted a means to talk about people without referring to them by their real names. Thus, they devised a list of bird names that served as a very simple code mechanism.

Puthoff, Jones, and I were given Partridge, Hummingbird, and Chickadee respectively. At least that's what I've been told. To be honest, I have trouble remembering if that is the correct order, though I have seen it in other publications. It was not terribly important at the time, nor did it ever impact any of my research.

The vast majority of the material on the Internet concerning the mystical Aviary is simply amusing. Several people have stated I was known as the Penguin, and even published articles with that title. Others, equally unknowledgeable, have said that I actually run this cabal. All of that information is both false and irrelevant.

The National Archives

All FOIA requests for dated presidential documents are sent to the National Archives for processing as they are the custodians for the White House. What is truly amazing is that there are two sets of documents that are highlighted on the National Archives Web site. One relates to the assassination of President John F. Kennedy and the other is Majestic 12. That alone should alert the reader as to just how widespread the MJ-12 mystery has become.

One of the key documents that proponents point to as proof of authenticity is what is known as the Cutler-Twining Memo of July 14, 1954. The alleged memo was most conveniently found out of place in a file box at the National Archives by Jamie Shandera. The memo purports to be from Robert Cutler, then a special assistant to the President, and directs General Nathan Twining, then the Chief of Staff of the U.S. Air Force, that an MJ-12 briefing will be added to the previously scheduled White House meeting on July 16, 1954. There is no other mention of the subject to be covered as it is assumed that those involved were already cognizant. Of course those steeped in the MJ-12 mystery assume that it refers to research on the Roswell crash.

One of the UFO researchers who studied the Majestic controversy in detail is nuclear physicist and well-known author of UFO books Stanton Friedman. As previously indicated, he wrote a book called *Top Secret/Majic* that is dedicated to this subject. In the book he explored the Cutler-Twining memo and concluded: "Thus it appears that the Cutler-Twining memo is the real thing, indicating that there was a Top-Secret MJ-12 group."

Unfortunately, the National Archives does not agree with his premise. They have posted a public response in the hope of forestalling additional requests for Majestic materials. Their comments in the entirety read:

"MAJESTIC 12" OR "MJ-12" REFERENCE REPORT
The National Archives has received many requests for documentation and information about "Project MJ-12." Many of the inquiries concern a

memorandum from Robert Cutler to Gen. Nathan Twining, dated July 14, 1954. This particular document poses problems for the following reasons:

1. The document was located in Record Group 341, entry 267. The series is filed by a Top Secret register number. This document does not bear such a number.

2. The document is filed in the folder T4–1846. There are no other documents in the folder regarding "NSC/MJ-12."

3. Researchers on the staff of the National Archives have searched in the records of the Secretary of Defense, the Joint Chiefs of Staff, Headquarters U.S. Air Force, and in other related files. No further information has been found on this subject.

4. Inquiries to the U.S. Air Force, the Joint Chiefs of Staff, and the National Security Council failed to produce further information.

5. The Freedom of Information Office of the National Security Council informed the National Archives that "Top Secret Restricted Information" is a marking which did not come into use at the National Security Council until the Nixon Administration. The Eisenhower Presidential Library also confirm that this particular marking was not used during the Eisenhower Administration.

6. The document in question does not bear an official government letterhead or watermark. The NARA conservation specialist examined the paper and determined it was a ribbon copy prepared on "diction onionskin." The Eisenhower Library has examined a representative sample of the documents in its collection of the Cutler papers. All documents in the sample created by Mr. Cutler while he served on the NSC staff have an eagle watermark in the bond paper. The onionskin carbon copies have either an eagle watermark or no watermark at all. Most documents sent out by the NSC were prepared on White House letterhead paper. For the brief period when Mr. Cutler left the NSC, his carbon copies were prepared on "prestige onionskin."

7. The National Archives searched the Official Meeting Minute Files of the National Security Council and found no record of a NSC meeting on July 16, 1954. A search of all NSC Meeting Minutes for July 1954 found no mention of MJ-12 nor Majestic.

8. The Judicial, Fiscal and Social Branch searched the indices of the NSC records and found no listing for: MJ-12, Majestic, unidentified flying objects, UFO, flying saucers, or flying discs.

9. NAJA found a memo in a folder titled "Special Meeting July 16, 1956" which indicated that NSC members would be called to a civil defense exercise on July 16, 1956.

10. The Eisenhower Library states, in a letter to the Military Reference Branch, dated July 16, 1987: "President Eisenhower's Appointment Books contain no entry for a special meeting on July 16, 1954 which might have included a briefing on MJ-12. Even when the President had 'off the record' meetings, the Appointment Books contain entries indicating the time of the meeting and the participants. . . .

"The Declassification office of the National Security Council has informed us that it has no record of any declassification action having been taken on this memorandum or any other documents on this alleged project . . ."

Robert Cutler, at the direction of President Eisenhower, was visiting overseas military installations on the day he supposedly issued this memorandum—July 14, 1954. The Administration Series in Eisenhower's Papers as President contains Cutler's memorandum and report to the President upon his return from the trip. The memorandum is dated July 20, 1954 and refers to Cutler's visits to installations in Europe and North Africa between July 3 and 15. Also, within the NSC Staff Papers is a memorandum dated July 3, 1954, from Cutler to his two subordinates, James S. Ia and J. Patrick Cone, explaining how they should handle NSC administrative matters during his absence; one would assume that if the memorandum to Twining were genuine, Lay or Cone would have signed it.

The National Archives responded to my recent FOIA request regarding MJ-12 with the same general language. Even after I specifically stated that UFOs were not the topic of my inquiry, the standard response was mailed. There may be another possibility.

COG: An Alternative Solution

What then could be the reality underlying Majestic 12? Is it possible the organization once existed and that a collection of influential people was brought together for some purpose? The answer may be yes. A reliable, vetted, and confidential source, who states he had access to MJ-12 material, indicated this was a real group. He also indicated that there would be no reports at the Department of Defense level as everything was controlled by the White House. However, he firmly acknowledges that the topics the group was involved in studying had nothing to do with the Roswell crash in particular or UFOs in general.

The following is speculation, but that clue caused me to think seriously about what such a body as MJ-12 might be involved in. It was Hal Puthoff who pointed me toward what could be the real answer—*Continuity of Government* or COG. For decades this was one of the most highly guarded secrets in America. Formally initiated under President Eisenhower at the height of the Cold War, COG was designed to prevent nuclear decapitation of the U.S. Government. It would appear that some of those plans remain classified and have been adapted to current counterterrorism circumstances. The point is that in those early post–World War II days, nerves were frayed, tensions were high, and a plan for national survival was needed.

We do know that continuation of leadership was a primary concern of President Truman. In 1945, only two months after being sworn into office following the death of President Franklin D. Roosevelt, he asked Congress to designate the line of succession. According to the *Congressional Research Service*, "He [Truman] noted that, in naming his Cabinet members, a President chose his successor, and concluded that, 'I do not believe that in a democracy this power should rest with the Chief Executive.'" Therefore, the idea that President Truman would make continuity of government a primary focus is fundamentally sound.

Very important from an MJ-12 perspective, all of the pieces fit, including timing, mission, and membership. On July 26, 1947, President Harry S. Truman signed the National Security Act that realigned and reorganized the U.S. Armed Forces, foreign policy, and Intelligence Community apparatus. It was the National Security Act that created the Air Force as a separate and equal branch of the Department of Defense. The Act also established the National Security Council as a centralized body for coordination of national security policy within the Executive Branch. In addition it created the first peacetime intelligence organization, the Central Intelligence Agency. Notably, this act did not go into effect until September 18 of that year and one day after James Forrestal was confirmed by the U.S. Senate as the first Secretary of Defense. The memo from Truman to Forrestal, directing him to initiate Operation Majestic 12 "with all due speed and caution," is signed just six days later (September 24, 1947). While this memo has not been authenticated, the content is commensurate with activities that followed.

It was under the COG plan that hardened secret underground bases such as Mount Weather in western Virginia were constructed. A few years ago it was revealed that a swanky resort called Greenbrier, in White Sulphur Springs, West Virginia, actually had hidden subterranean facilities and was the place that Members of Congress would be sequestered in case of a nuclear exchange. Amazingly, it is now open to tourists.

However, complex plans, such as COG, do not materialize out of thin air. They require extensive thought and careful planning. The alleged composition of MJ-12 was exactly right for the task of developing a plan to safeguard American leadership. They had the brainpower and experience to tackle such a problem. Further, creating a body of senior advisors was the normal manner by which government agencies approached complex issues such as restoring duly authorized leadership under catastrophic circumstances. In fact, that process of appointing advisors continues to be a widely used norm. Membership of such panels is usually directly related to the level of the office establishing the study. Those named as the MJ-12 constituency dovetails appropriately with a body created that might advise a POTUS. The dearth of written substantiation is also reasonable. Other extremely sensitive projects were known to be conducted with little, or no, paper trail.

In Stan Friedman's book he goes to great lengths to attempt to explain why Donald Menzel, a harsh UFO critic, would be included on a panel with direct access to material proving his position to be wrong. Stan concluded that it was because Menzel was leading a double life. While he publicly denounced UFOs, Stan contends Menzel was a closet insider and knew that he was intentionally misleading the American people. A much simpler answer would be that Menzel was a member of a group addressing pressing problems of strategic importance that were not related to UFOs. Certainly COG is a perfect fit. They were the right people, at the right time, involved in the right mission. That is, if they ever did exist.

Summary

The MJ-12 myth eclipses all other UFO-related conspiracies and cannot be ignored if one is to engage in research in this field. As noted, the notion has even permeated the National Archives and invokes images of a supersecret cabal that seems near omniscient. While no other organization in the world could function with such nefarious efficiency, believers attribute mystical powers to these keepers of the flame. Having roots deep in the human psyche, this obvious contradiction in capabilities fails to extinguish the deeply held belief that a magical federation controls all aspects of official knowledge regarding UFOs.

It is certain that people will go to great lengths to deceive others, even if there is little tangible to be gained from the endeavor. This is dramatically demonstrated by the plethora of bogus documents that have been discovered. The MJ-12 stories are part of a great game that feeds back on itself and, once ignited, leads into perpetuity. While there is no cabal that relentlessly squashes curious UFO investigators, such a notion is useful in explaining why bad things happen, especially when one's lifestyle entices danger. The bogeyman still holds a vaunted place in modern society.

There is serious doubt as to whether or not there ever was an organization known as Majestic or MJ-12. However, if such an entity did exist, it had nothing to do with UFOs.

Of great importance was the Apollo program, our manned mission to the moon. Covered in the next chapter is what the lunar astronauts were or were not told about UFOs. That is extremely significant.

8

THE APOLLO LUNAR PROGRAM

One small step for man, one giant leap for mankind
—NEIL ARMSTRONG. APOLLO 11

The Mission

"First, I believe that this nation should commit itself to achieving the goal, before this decade is out, of landing a man on the moon and returning him safely to the earth." Those profound words were spoken by President John F. Kennedy in his Special Message to the Congress on Urgent National Needs on May 25, 1961. They set in motion a program that would change forever how we view space—but this speech was about far more than going to the moon. The moon was important, but it was a means to achieve a much larger goal—technological superiority in the world and to counter the spread of communism everywhere.

In that speech President Kennedy went on to say, "No single space project in this period will be more impressive to mankind, or more important for the long-range exploration of space; and none will be so difficult or expensive to accomplish. We propose to accelerate the development of the appropriate lunar space craft. We propose to develop alternate liquid and solid fuel boosters, much larger than any now being developed, until certain which is superior. We propose additional funds for other engine development and for unmanned explorations—explorations which are particularly important for one purpose which this nation will never overlook: the survival of the man who first makes this daring flight. But in a very real sense, it will not be one man going to the Moon—if we make this judgment affirmatively; it will be an entire nation. For all of us must work to put him there."

Apollo Was About Far More than Visiting the Lunar Surface

Now, nearly half a century after the lunar quest began, most Americans are blissfully unaware of the events that led up to this monumental decision. Four years before the speech, in 1957, the world had been shocked by the launch of

Sputnik by the USSR. The space race had already begun in earnest. Then, on April 12, 1961, cosmonaut Yuri Gagarin had become the first human in space. This was a great embarrassment for the United States. On May 5, 1961, Alan Shepard became the first American to enter space; however, he only flew on a short suborbital flight instead of orbiting Earth, as Gagarin had done.

There were other significant pressures on Kennedy as well. The invasion of Cuba at the Bay of Pigs had been botched in mid-April, and he was in no small part a contributing factor. The Cold War was in full swing and tensions were increasing in every sector and Berlin would soon become a severe test of wills once again. Dealings with Russian Premier Nikita Khrushchev was not going well and just over a year later the Cuban Missile Crisis would bring the world to the brink of total war. Feeling the political pinch, Kennedy wanted to announce a program with goals that the United States had a strong chance of achieving before the Soviet Union.

While the President decided on the Moon as an objective, he then had to set about influencing the American public to embrace this cause. On September 12, 1962, he gave a famous speech at Rice University Stadium in Houston, Texas, in which he further articulated the issues. Kennedy said, "We choose to go to the moon. We choose to go to the moon in this decade and do the other things, not because they are easy, but because they are hard, because that goal will serve to organize and measure the best of our energies and skills, because that challenge is one that we are willing to accept, one we are unwilling to postpone, and one which we intend to win, and the others, too. It is for these reasons that I regard the decision last year to shift our efforts in space from low to high gear as among the most important decisions that will be made during my incumbency in the office of the Presidency."

Of course, one of the major challenges was paying for this venture. For American publicly funded projects only the construction of the Panama Canal in modern peacetime and the Manhattan Project in war were comparable in scope. At the present time NASA's budget has been severely cut back and plans to return to the Moon as a government program scrapped, so it is hard to imagine the economic commitment required to meet Kennedy's goals. At its height in spending in 1966, NASA had a budget that was 5.5 percent of all Federal spending. By comparison, in 2010 NASA's proposed budget is 0.52 percent on the Federal budget. That translates to the Apollo program having about ten times the financial support that NASA has at the moment. The Apollo program was accomplished employing 500,000 workers in 20,000 companies. The relative importance of the Apollo mission takes on additional significance when the other government priorities are considered. Most of the

Vietnam War was during this period. Wars are always expensive, yet they found money for space. The civil rights movements were extremely active during that time, and also cost the taxpayers' considerable sums. Despite all that, Apollo remained well funded through three presidents.

While President Kennedy gets the credit for the Apollo mission, he did not think up the lunar landing program in a vacuum. The concept had been drawn up under President Dwight D. Eisenhower. Worth noting is that General Trudeau, who was mentioned in Chapter 2, and U.S. Army experts, along with their German counterparts, had drawn up Project Horizon to establish bases on the Moon and had considered interplanetary defense issues. There were many intermediate steps that had to be completed. Starting in 1959 there was the single astronaut Mercury program with six manned flights. That was followed by Gemini, transporting two astronauts in earth orbit on ten missions. Then, prior to sending humans to the Moon, there were a series of probes for data gathering that preceded them. Going to the Moon was a complex, very detailed plan, and it is believed that it was the organizational process that allowed the United States to beat the Soviets in this arena.

On July 20, 1969, about two years ahead of the deadline proposed by President Kennedy, the crew of Apollo 11 sent that momentous message, *"Tranquility Base here, the Eagle has landed."* Only a few hours later, with the entire world watching on television, Lunar Astronaut Neil Armstrong uttered the most famous phrase in extraterrestrial history, *"That's one small step for man, one giant leap for mankind."* Without a doubt, America had unequivocally demonstrated its global technological dominance. Finally Sputnik was behind us.

The Lunar Astronauts Did Not Have a Contingency Plan for Meeting ET

By now some readers are wondering what the history of the Apollo program has to do with the study of UFOs and ETs. Actually quite a lot. Set aside the bogus notion that the whole lunar expedition was faked and filmed on a movie studio set. Man went to the Moon! But there have been many claims about what the lunar astronauts saw or didn't see. There are still claims of bases for ET on the back side of the Moon. Conspiracy theorists have no end of stories about all of these hypotheses. However, the Apollo program lays bare the fact that the U.S. Government was not hiding any previous evidence concerning the existence of extraterrestrial visitors.

As noted, going to the Moon was a strategic move with global implications that would last for decades. While Americans did push the envelope when it came to technological advancement, everything humanly possible was done to minimize risk and eliminate surprises. Even at that, there were serious accidents. On January 27, 1967, Apollo 1, as it was retroactively designated, caught fire on the pad during a training exercise. Due to design flaws with the hatch, the astronauts could not escape and the onboard crew—Virgil "Gus" Grissom, Ed White, and Roger Chaffee—perished. The fate of Apollo 13, which suffered catastrophic failure en route to the Moon, is well known. Equally documented is the remarkable response by the NASA teams on the ground that worked feverishly to find solutions, and eventually returned the crew safely to Earth. In order for that effort to be undertaken, huge amounts of data were shared. These were drawn from existing databases that were available to the technicians. They did not have to acquire new information from secret sources to accomplish their mission.

The reason that the Apollo missions, including Apollo 13, were successful in lunar exploration was that they had contingency plans for everything imaginable. When the unimagined happened, the information required to solve the problem was widely available to all the engineers and technicians who needed it. It seems inconceivable that the astronauts would not have a contingency plan if NASA thought there was the remotest possibility that they might encounter ET.

What about ET?

Now to address ET directly. Those old enough to have observed Apollo 11 can remember the return of the spacecraft. After recovery from the Pacific Ocean, the crew members immediately donned Biological Isolation Garments. They were then transported to a waiting Mobile Quarantine Facility located on the waiting aircraft carrier, the USS *Hornet*. The returning Apollo astronauts were taken in that chamber first to Pearl Harbor and Hickam Air Force Base in Hawaii, and then flown to Ellington Air Force Base, Texas, where they remained in medical isolation for a total of 21 days. The reason for the extreme precautions was concern that while on the Moon, microorganisms foreign to Earth, and potentially dangerous to humans, might have been transported back with them. The lunar samples were also subjected to isolation. Yes, NASA was well aware of the severe medical problems that have occurred when cultures with developed immunity infect another vulnerable society during initial contact. These procedures were followed for Apollo 11, Apollo 12, and Apollo 14 before being dropped. Apollo 13 would have experienced similar isolation but the astronauts never touched the lunar surface.

The point is that NASA was concerned about the remote possibility that the lunar astronauts might pick up microbes on the Moon. However, NASA had no contingency plans for the possibility that the lunar astronauts might encounter an extraterrestrial being from an advanced civilization, one capable of visiting the Earth or have bases on the back side of the Moon. If there was knowledge that such an encounter might have occurred, or it was even considered a slight possibility, appropriate contingency plans would have been made known, at least to the key people who would have to handle the situation. If considerable knowledge of the advanced society was available to government scientists, protocols, like a secret handshake or dinner invitations at the White House, would have been included in the briefings.

No such contingency plans or procedures were included in NASA's preparation. The Apollo missions took place more than twenty years after the Roswell incident with its reports of alien beings, either alive or dead, depending on the story. Would not the mere confirmed knowledge that aliens do travel in space and may have visited Earth, or the Moon, warrant contingency plans? It is inconceivable that no such warnings were given had that knowledge been available anywhere within the U.S. Government. Remember, it was our global position that was at stake. Any and all precautions would have been taken. They were not, which provides a very solid answer to this issue.

Meetings with Lunar Astronauts

Over the years I have met with three of the lunar astronauts. Both Dr. Harrison "Jack" Schmitt and Dr. Edgar Mitchell were on the NIDS Science Advisory Board (SAB) with me where topics of phenomenology were discussed openly. As he speaks at their conferences frequently, Edgar Mitchell is well known to Ufologists. However, Jack Schmitt has not addressed any of those groups, nor is he likely to in the future.

Harrison Schmitt

While still working at Los Alamos and prior to the formation of NIDS, Schmitt and I met in Albuquerque, New Mexico, to discuss UFOs. Schmitt came to the astronaut corps as a scientist with a doctorate in geology from Harvard. Before joining NASA, he was with the U.S. Geological Survey's Astrogeology Center at Flagstaff, Arizona. He was project chief for lunar field geological methods and participated in photo and telescopic mapping of the Moon. Schmitt occupied the lunar module pilot seat for Apollo 17 and holds the record for the longest lunar surface extravehicular activities (twenty-two hours, four minutes). He was also the last man on the Moon of the Apollo project.

After leaving NASA he successfully ran for the U.S. Senate where he served one term representing his home state of New Mexico. His openness to exploring phenomena was attested to when he served as a chairman on a conference to look into the cattle mutilations that were a problem in that state. While there was no doubt that cattle were found dead and mutilated, the cause was (and remains) unknown. The findings of that conference were problematic as they certainly did not cover all of the facts involved in the incidents. Nonetheless, he did step forward and publicly address a contentious issue.

In our talk in Albuquerque Schmitt was clear that he was very skeptical of the claims made about UFOs. Still, he was prepared to listen to the details of the Advanced Theoretical Physics group's exploration of the topic. Like all of the other astronauts who have ventured around the Moon, he was quite sure there were no secret ET bases lurking on the back side of that body. As we got into our cars to leave, Schmitt came back over to me to ask a classic question about the motives an intelligent extraterrestrial exploration unit might have for behaving as they appear to. Understanding the motives of ET is well beyond my grasp. But at least Jack Schmitt was open to thinking about the issues in a serious manner.

Buzz Aldrin

Aldrin went to the United States Military Academy at West Point, New York, graduating third in his class, and then earned a doctorate of science in Astronautics from Massachusetts Institute of Technology. Aldrin entered the U.S. Air Force and was a combat pilot during the Korean War where he was credited with downing two enemy MIG 15 fighters. While his partner on Apollo 11 and first man on the moon, Neil Armstrong, generally stayed out of the eye after leaving NASA, Aldrin has been highly visible and leading the charge in advocating various space ventures.

Buzz Aldrin and I have met on several occasions but rarely touched on UFOs. That is clearly not his favorite topic, and unlike Schmitt and Mitchell, he is fairly hostile on the subject. There is controversy among UFO advocates concerning Aldrin's remarks about a possible sighting during the Apollo 11 mission. While proponents claim that this was a real encounter, Aldrin states he is 99.9 percent sure the object was a piece of the craft that had become detached.

The first meetings with Aldrin were brief and happened at Burt Rutan's rollouts at Burt's company, Scaled Composites, in Mojave, California. That included the time when Rutan first publicly announced his intention to put a civilian into space with *SpaceShipOne*. In 2008 we spent more time together. With our wives, we were guests at Tom Clancy's birthday party aboard the

USS *New Jersey* and had an opportunity for discussion of various topics. The issue of UFOs, which I brought up, was short-lived. Aldrin's public position appears to be exactly the same as his private one—highly skeptical.

Edgar Mitchell

Dr. Edgar Mitchell I consider both a personal friend and a great American hero. After college Edgar entered the U.S. Navy where he flew off aircraft carriers, one of the most demanding and dangerous jobs a pilot can have. He later obtained a Doctorate of Science degree in Aeronautics and Astronautics from the Massachusetts Institute of Technology and as a navy captain joined NASA. He was the lunar module pilot for Apollo 14 and became the sixth man to walk on the Moon. Among our discussions has been the tension in preparation for that momentous mission. Following the ill-fated Apollo 13, his was the first flight after something had gone seriously wrong in space. The intestinal fortitude it took for the crew of Apollo 14 speaks volumes about the character of these men.

As mentioned, Edgar was on the NIDS SAB but in addition to those meetings we have had numerous discussions specifically about the topic. Some have taken place at our respective homes and other talks have been at various conventions. We have some different data points, and he has spoken publicly in support of the UFO issue in general and Roswell in particular. Here, while I respect his opinion, we agree to disagree. However, in our talks Dr. Mitchell is very explicit that none of his information about UFOs or possible extraterrestrials visiting Earth came from NASA. In fact, after his return from the Apollo 14 mission, he was in charge of security for the additional scheduled flights. He is clear that there was nothing related to security measures for NASA that in any way suggested an encounter with advanced sentient beings was even a remote possibility. All of his UFO-related material comes from sources external to NASA.

The important factor is that all the astronauts agree that they never had a NASA briefing regarding the likelihood of encountering an ET while involved in Apollo missions. Both Schmitt and Aldrin are openly skeptical about the UFO topic, which would support the fact that NASA had no special knowledge about ETs.

In fact NASA, as an organization, remains extremely hostile to the UFO topic in general. On the official NASA Web site, astrobiologist and Senior Scientist David Morrison answered a student's question about UFOs as follows: "No, I can't imagine that NASA would have any problems with something that doesn't exist. The sad thing about UFO reports is that they distract so many people from understanding and enjoying real science. There is a lot of nonsense on the Internet about UFOs." The official NASA position is clearly in line with the Condon Report, even to the point of lamenting the adverse impact on education.

OMG ET Is Coming

Imagine receiving that as a text message! Of course, the public knows from every existing movie line or television story related to ET that the evil governments would bar the public from learning about the existence of an advanced extraterrestrial civilization. Any contact would result in the military swooping in, most likely in biological protective gear, and placing everyone in quarantine. In *Close Encounters of the Third Kind*, with preparations for contact well under way, unsuspecting tourists were sprayed with some magical incapacitating gas. In many other UFO movies the good civilians must save themselves or humanity from both the extraterrestrials and the actions of their government. The list is long and would include *Contact, Independence Day, The Abyss, The Day the Earth Stood Still*, and *War of the Worlds*. Even the cute ET in *E.T.: The Extra-Terrestrial* was ruthlessly pursued by government agents, only to be saved by innocent children. In most of these stories the ETs, even when masquerading as friends, are really out to do us harm, as in the *Twilight Zone* episode "To Serve Man." Given those scenarios, why wouldn't NASA keep contact secret?

The reality is very different as I learned from a friend of mine, Bob Kupperman. We met at Los Alamos where he was a fellow. Most of his time was spent in Washington where he worked at the Center for Strategic and International Studies (CSIS). Dr. Kupperman was a terrorism expert and one of the first experts to sound the alarm about the dangers of biological weapons. His studies were a couple of decades before his time; it was difficult to get anyone to listen. In fact, in his obituary it was stated that if people had paid attention to him, 9/11 might not have occurred. Kupperman served the nation in several key positions. He served as the Chief Scientist of the Arms Control and Disarmament Agency, the Executive Director of the Office of Emergency Preparedness, and transition director at the Federal Emergency Management Agency better known as FEMA.

Our relationship began as he was a Washington insider who supported my efforts in nonlethal weapons. Over time we met at his CSIS office, at LANL and at my home in Santa Fe. We had opportunities to discuss a wide range of topics, one of which included UFOs. While Kupperman was open to discussions about the topic, I clearly knew more about UFOs than he did. Remember, Bob was a guy who had been involved with the nation's top organizations that dealt with emergencies. The responsibilities of FEMA went far beyond natural disaster relief and the entire scope would amaze most people. In our discussions, it was obvious that contingencies for even contact with ET were not on its priority list—it was never even thought about in the emergency organizations at which he worked.

Unfortunately Kupperman had Parkinson's disease from which he died in 2006. When I visited Washington, and he was healthy enough, we would go to dinner with his first wife, Helen. It is Helen Kupperman who really takes the lead in this part of the chapter, for she was an attorney at NASA and familiar with the issue of potential for contact with an extraterrestrial civilization. She told me about the meetings that had been held regarding the role of NASA in the event that contact with ET was made. She noted that the topic had been discussed in some detail, but from a purely hypothetical standpoint. Even so, the topic was given serious consideration. The deliberations eventually made their way into international legal documents.

Immediate Disclosure Is Required

Contrary to Hollywood hype, there are binding treaties that call for immediate worldwide public release of information should the existence of ET be confirmed. Helen Kupperman was kind enough to give me her copy of the book that covers international space law. The specific reference can be found in the *Treaty on Principles Governing the Activities of States in the Exploration and Use of Outer Space, Including the Moon and Other Celestial Bodies*. For the United States it was Proclaimed by the President on October 10, 1967, and was entered into force the same day. Signatories to this treaty include nearly every country on the planet.

There are numerous issues addressed in this treaty, including prohibition of nuclear weapons, rescue and recovery of astronauts, freedom of exploration, liability for space activities, and notification of space activities. Article XI is applicable and states, " In order to promote international co-operation in the peaceful exploration and use of outer space, States Parties to the Treaty conducting activities in outer space, including the moon and other celestial bodies, agree to inform the Secretary-General of the United Nations as well as the public and international scientific community, to the greatest extent feasible and practicable, of the nature, conduct, locations and results of such activities. *On receiving the said information, the Secretary-General of the United Nations should prepare to disseminate it immediately and effectively*" (emphasis mine). As a NASA lawyer, Helen Kupperman thought that this passage, as well as the remainer of the treaty, bound the United States to the release of information upon confirmation of ETs existence.

Note that this treaty was in force at the time of the Apollo missions and our lunar astronauts would be required to report any contact that occurred while in space. I suppose that one could argue that if ET made contact with people while on Earth, that information would not be bound by the treaty. However, any reaction we might have that included space activities would be covered by the treaty. Bottom line: *If ET calls, the U.S. Government is legally bound to immediate dissemination of that news to the entire world.*

Summary

There is a great deal of misinformation about the Apollo missions and UFOs. While several Apollo astronauts claim to have seen UFOs, none of those sightings that remain unexplained were during the Apollo flights. The Apollo program was far more important than just going to the Moon—it was about technological dominance of the entire world. Nothing, repeat nothing, that could be contemplated was left to chance. There were contingency plans for every conceivable exigency. Yet, none of those included what to say to ETs if they were encountered. That alone clearly spells out what the government knew about the topic.

While NASA remains unaware of any contact with ETs, there are people who believe that a UFO has been reverse-engineered. The following chapter examines that hypothesis.

9

REVERSE-ENGINEERING UFOS

Reality is only an illusion, albeit a very persistent one.
—ALBERT EINSTEIN

A prevalent theme of conspiracy theorists is that either secret elements of the U.S. Government, and/or the unidentified *THEY*, have successfully reverse-engineered UFOs. Usually called alien reproduction vehicles (ARV), the lore has these crafts operational, under human control, and flying out of Groom Lake near Area 51. Exactly where the crash material came from is not clear, but most people assume it was retrieved from a crash near Roswell, New Mexico, in 1947.

Not One, But Many Crashes Alleged

Those steeped in the wrecked UFO mystique have heard of a considerable number of crash incidents at sites around the world. Besides Roswell, and taken from multiple sources, some of the more infamous recorded events include:

— 1897: Aurora, Texas. Craft crashed into a windmill and exploded, possible alien burial.
— 1908: Tunguska, Siberia. Conventional theory is this was a meteor, but some believe a UFO crashed.
— 1941: Cape Girardeau, Missouri. Crash with dead alien bodies.
— 1947: Maury Island, Tacoma. Slag material dumped on beach.
— 1947: Paradise Valley, Arizona. Crash with two humanoid bodies.
— 1948: Aztec, New Mexico. Crash with 2 to 16 bodies (depending on the source).
— 1952: Spitzbergen, Norway. A 150-meter diameter disc crashed.

— 1953: Kingman, Arizona. A 30-foot-diameter disc crashed
   with bodies.
— 1953: Fort Polk, Louisiana. Egg-shaped UFO, one dead and
   three living aliens.
— 1957: Ubatuba, Brazil. Craft almost crashed into the sea, then
   exploded.
— 1965: Kecksburg, Pennsylvania. Military recovered some
   form of crashed object.
— 1965: Pretoria, South Africa. Sounds more like a landing
   than a crash.
— 1965 Congo. UFO exploded and debris recovered.
— 1967: Shag Harbor, Nova Scotia. Object crashed into the sea,
   nothing recovered.
— 1974: Llandrillo, Wales, UK. Disc sighted falling over city of
   Clwyd.
— 1974: Chihuahua, Mexico. Radar tracked falling object that
   disappeared.
— 1989: Sverdlovsk, Russia. Crash with alien bodies.
— 1996: Central Israel. Fiery objects hit a stationary orb and
   exploded.

This is not a complete list but the contention that there were so many possible
UFO crashes is puzzling. In November 2008 George Noory was broadcasting
on his extremely popular *Coast to Coast am* late-night radio talk show from Las
Vegas when I made the comment: *"It appears that ET must have a quality control
problem."* According to the proponents of these crashed UFO theories, there is
an alien civilization, possibly several, which has developed the technology to
traverse the universe, visit Earth on demand, and yet they keep falling down.
Since some people believe that the Roswell incident happened when two UFOs
ran into each other, and UFOs reportedly have hit aircraft on more than one
occasion, it would mean that ET has not perfected their collision avoidance systems either. For civilizations that are ostensibly at least a thousand years in advance of our own, it seems surprising that some of their subsystems are not even
on par with the current aircraft and automotive technologies of us Earthlings.

The Craft Flies

For proof that someone has re-created a UFO, proponents often point to a videotape that was allegedly smuggled out of the Nellis Air Force Base Test Range
by a rogue government worker. Since the tape would have been illegally

obtained, the source remained unidentified when they sold it to commercial television. It was played on both *Hard Copy* and *Sightings* in 1995. If the object is under human control at that location, then it could only belong to the U.S. Government. It does not, however, confirm that it was built by Americans.

For many years even the existence of Area 51 was not officially acknowledged by the Air Force. That policy continued even after satellite photos were available to the public that clearly depicted long runways and many maintenance facilities. It is only in the past couple of years that any official statements have been authorized. One of the major functions of this remote, closed facility has been to test foreign systems in order to exploit vulnerabilities.

Regarding the exploited videotape, there are serious problems with the notion that a black project would have been flown in front of the cameras that are known to be located in the area. The transcript of the controllers indicates that they were surprised by the object and did not know its origin. To be sure, strange things do fly in the Nevada desert. This is the location where some of the most secret aircraft have been developed and tested. However, a test flight in broad daylight being accidentally caught by cameras is extremely unlikely. The capabilities and locations of all of the cameras were known and care was taken to avoid inadvertent disclosure of tests. Also unlikely is the idea that some contractor would risk the rest of his or her life behind bars to steal the tape and sell it to commercial television. After all, there would have been far higher bidders available. Adversarial countries or companies would pay top dollar for information leading to development of such exotic technology. From the seller's perspective, having it shown publicly would dramatically increase the probability of his or her long-term incarceration. Remember, there would only be a very small number of suspects, all of whom would have been under direct control of the government, and for whom taking polygraph tests would not be optional. Nonetheless, this is the evidence some proponents point to as proof of reverse engineering.

Dr. Steven Greer with his *Disclosure Project* has been one of the leading proponents stating that such reverse engineering has been accomplished. The following is a quote taken from the *"Special Presidential Briefing for President Barack Obama"* that was posted on their Web site: *"Alien Reproduction Vehicles (ARVs)—these are advanced anti-gravity aircraft that have been fully operational since at least the late 1950s to early 1960s. Many so-called UFO reports by civilians and military personnel are of such ARVs. They constitute an unacknowledged or 'black' Air Force and these ARVs are capable of extraordinary speed, maneuverability and lift/hover. By 2009, these technologies had gone through many genera-*

tions of refinement and, if deployed, could easily hoax or simulate an Extraterrestrial Vehicle (ETV). (Note that a UFO is a nonspecific term and could be either an ARV or an ETV.)" (Web site accessed by author, February 16, 2010.)

Other writers concur with this conspiracy. In *Need to Know,* British journalist Timothy Good wrote that based on "information via a high-ranking source . . . a number of alien craft have been recovered and elements of the military and Intelligence Community have been engaged in developing, *with alien assistance,* the highly advanced propulsion systems that power these craft" (emphasis added). In Good's model, not only have *THEY* reverse-engineered UFOs, but extraterrestrials actually participated in the process.

There Is Interest in Hypersonic Flight

Based on the foregoing, the proponents of these conspiracy theories believe that *THEY* have been withholding critical technologies for more than half a century. In one case they specifically point to the U.S. Air Force. It is interesting to note that every few years the Air Force Science Advisory Board (SAB) has been asked to explore aircraft that can fly extremely fast—Mach 9 or above. Their 2000 study, *Report on Why and Whither Hypersonics Research in the U.S. Air Force, SAB-TR-00-03,* is not classified and can be viewed on the Internet. The conclusion is that you can't get there from here. In addition, one of the Air Force SAB members did some calculations for me. Using hydrocarbon fuels, as they now do, the weight of the aircraft would be a million pounds and must fly at altitudes above 95 kilometers and below 110 kilometers. Below that level the drag on the aircraft is too high, while above 110 kilometers there is not sufficient oxygen for combustion. The minimum distance flown would be 1,200 miles. The analysis suggests this is not a militarily useful aircraft. If the ARV hypothesis is believed, then we must assume that the Air Force SAB doesn't have contact with the Air Force mentioned in the memo.

Of course the proponents will claim all of these activities are part of a cover story. When we consider the amount of money the Department of Defense has spent on researching advanced aerodynamics in the development of new aircraft over the past five decades, and that these secret capabilities have not been deployed during the various wars we have fought, the evidence for the existence of ARV disintegrates. Further, sequestering such technology, while squandering tens to hundreds of billions of dollars on already antiquated projects or allowing unnecessary American casualties, would not be just gross malfeasance; I would argue it would be treasonous.

It's About Energy, Not Widgets Flying

The assumption that the U.S. Government has reverse engineered an alien spacecraft exposes an even larger problem of faulty logic. The most important aspect of UFOs would not be that unusual quixotic objects fly around. It would be the propulsion system that would be the key to far greater uses than just aviation. For a UFO to function as has been described, then *THEY* must have developed an advanced energy system. Making objects fly would be but a relatively minor function of that amazing discovery. The real application would be in providing energy across a broad spectrum of needs. It would be geotransformational nature. As I postulated in *Winning the War* (St. Martin's Press, 2003), development of a large-scale, moderate cost energy source that does not rely on petrochemicals would change global political dynamics. If that energy source can be made portable, as in any vehicle, then the importance would increase exponentially. Having studied the issue of transitioning a new energy system for many years, it is also known that it would take two to three decades to incorporate a dramatically different technology into our society.

The problem of integration is quite simple. If one has a pound of unobtainium, which contains sufficient energy to run a car for a thousand miles, it is useless unless there is a mechanism in the car to convert the energy from unobtainium into a force that makes the car move. The energy conversion process is now done, in most cases, with your internal combustion engine. Somehow the energy must be transmitted though the axle to the wheels. Once the new mechanism to make the vehicle move is engineered and installed, then there must be an energy distribution system. Energy distribution can be thought of as the power lines that provide electricity, filling stations that dispense petroleum fuels, even pipes that bring natural gas for home heating. All of these systems have a number of features in common; they are complex, dispersed, pervasive, expensive, and require intensive management. It has taken many decades for the current systems to be established and they would not be easily replaced by any alternative system. If there is any doubt, consider the fate of all-electric cars that were leased out for a short period of time. Aerospace engineer Burt Rutan allowed me to drive his leased EV1 over a decade ago. Battery powered, the vehicle was incredibly responsive and great for local driving. However, rather than expanding the market, General Motors withdrew all of the cars and destroyed them.

For the record, unobtainium does not exist. This is a notional term that allows planners to conceptualize how a unique new technology system might work. In this situation unobtainium is used as a metaphor to describe how the energy system that was allegedly developed from reverse engineering a UFO would have to be constructed.

Interestingly, the proponents of the UFO-developed energy hypothesis claim that it is the existing energy suppliers who are most threatened by the prospect of an alternative source being made widely available. They also claim that those that have developed this alternative energy system have done so on the basis of greed. Hypothetically, when the time was right, *THEY* would then introduce the new energy system and make incalculable amounts of profit.

There seems little doubt that petrochemical energy prices are manipulated by OPEC and this has had a major effect on global economics. Several times in recent decades the Arab oil producers have allowed the price of a barrel of crude to jump dramatically. In July 2008 the price of gasoline capped in the United States at just more than four dollars per gallon. Shortly thereafter the prices declined quickly and significantly. The reason was quite simple. OPEC had exceeded the pain threshold of most Americans. Due to price, people started to change their driving habits and gas-guzzling cars fell out of favor. In some extreme instances, owners actually torched their SUVs hoping the insurance money would allow them to buy a more economical car. The concern for OPEC was that if the pain continued, American scientists would get serious about developing alternative energy sources. Combined with the climate change phenomena, this experience did set in motion a green energy movement. That said, gas is still relatively cheap in the United States, particularly when compared to Europe, and we continue to use far more than our share based on our percentage of the world population.

Disruption of global macroeconomics is given as one of the problems with introduction of UFO-based energy technology. There was a stated need to maintain the stability of the current economic system and thus those holding back the technology had to understand the impact across the world. In 2009 the world saw the impact of mismanagement of large-scale economics, but it had nothing to do with ETs. Everyone should be aware of the meltdown and the extreme measures that were taken by the U.S. Government to stabilize the very rich. In addition, the volatility of energy processes, especially oil during the last half of the first decade of the twenty-first century, contributed greatly to global economic instability. Those problems were not isolated to the United States and countries around the world were forced to take drastic action. Countries that relied heavily on petroleum export for income were among the heaviest hit.

When examined in any detail, it should be apparent that one of two things happened that are directly related to the alleged UFO energy technology. The more complex is that those holding the technology totally miscalculated the economic system of the world, and were too late in allowing the release of the technology. To be successful, *THEY* must introduce the alternative energy system over a period of at least two decades. If we are to believe the proponents of the theory that reverse engineering has been accomplished, then *THEY* should have initiated the new system some time ago. In that way *THEY* would have capitalized on the new market and made the great amount of money hypothesized.

But *THEY* did not intervene. So the second, and far more rational, hypothesis is that *THEY* either don't exist, or at a minimum don't have access to UFO-developed energy technology. The macroeconomic reality is that at least ten trillion dollars (that's the *T* word, not the *B* word), possibly more, in real wealth simply disappeared from the global economy. Now, is it logical to assume that some organization, with a stated motive of making vast amounts of money, would allow ten trillion dollars to evaporate? The short answer is no!

The Challenge

That *THEY* would intentionally allow loss of wealth at that magnitude seems illogical, at least from my perspective. Ten trillion dollars is simply too much money to allow to disappear when there is an alternative immediately at hand. Therefore, my ten-trillion-dollar challenge was published. The purpose was to point out the fatal flaws in the thinking that *THEY* had the alternative energy resolution in hand, and yet sat by and allowed the nation and the world to go to the brink of economic disaster. The following is the memorandum that is dated March 1, 2009.

THE TEN TRILLION DOLLAR CHALLENGE

For many years the proponents of the Alien Reproduction Vehicle (ARV) have suggested that some unidentified agency (*THEY*) has had control over a crashed UFO and were successful in reverse engineering that craft. According to this theory, *THEY* have controlled all access to information pertaining to ARV. Select versions of this story suggest that some, but not all, former presidents of the United States of America, have been allowed to have this knowledge. Exactly who makes the decisions about access to the information is never articulated, but the proponents are convinced that *THEY* exist. Speculation regarding control runs from some deep black governmental agency, to private industrial organizations, or even extragovernmental bodies which are totally devoid of oversight from either the Executive or Legislative branches of government.

Those not steeped in UFO conspiracy theory may not be aware that there exists a substantial body of published information that states that *THEY* have gained the scientific knowledge necessary to provide limitless energy at very low cost. Ostensibly, that advance was derived from the UFO crash at Roswell in 1947. According to this legend, once successfully reverse engineered, *THEY* intentionally blocked access to this knowledge in order to literally take over the world through domination of the macro socioeconomic structures. Support for these theories goes far beyond a few loose nuts and gains credibility through commentary by highly credentialed persons, such as a former Canadian Minister of Defense, active and retired military officers, and experienced researchers.

The evidence provided to support their conclusions of the existence of ARV involves a few, hard to substantiate, eyewitness accounts and some low-quality videotape of questionable origin that purports to show an object morphing in an unusual manner. These same rumors suggest that ARV is housed at Area 51, or S-4, which are under U.S. Government control in southern Nevada.

Of course, development of such a craft implies intimate knowledge of some propulsion mechanism that is unknown to the traditional scientific community. If this has been accomplished, then development of a flying craft is minuscule compared with the true value of this monumental scientific breakthrough. Successful development of ARV means that a previously unknown energy source not only has been discovered, but has been engineered. Obviously, ARV is but one variation of use. However interesting ARV might be, it pales in comparison to other applications.

The main focus would be on energy generation. The proponents of the ARV thesis acknowledge this, and state that the rationale for keeping the new technology secret is so that *THEY* can dominate the energy market and make trillions of dollars through ownership of this technology. For the sake of discussion, we can set aside the argument that *THEY* have any legal jurisdiction over the application of such a breakthrough. Since the technology would have been derived, directly or indirectly, from U.S. Government–owned sources, the technology would belong rightfully to the American people.

However, since ARV does not exist, that position is moot. Prior to the 2008 national elections I stated publicly that if ARV were real, *THEY* would have revealed the existence of it in order to sway the elections in favor of the Republicans. As we all know, that revelation did not happen. Now there is another fact that has come to light that certainly argues most strongly against ARV. That is the decline in global wealth from 2007 through 2008. According to a report in *Business Week*, during that period there was a real decline in global wealth amounting to about ten trillion dollars. While unstable financial situations played a significant role, at the heart of the loss was the volatility in global energy sources. To believe that *THEY* control ARV, means that *THEY* were willing to allow an 8 percent decline in global wealth, while *THEY* had the capability to mitigate, or even reverse, this disastrous economic collapse. This is completely illogical and runs counter to the thesis that *THEY* were driven by greed and would use ARV technology to further their own economic interests. In fact, during the economic recession of 2008 (and continuing in 2009) more wealth was lost than *THEY* projected to make through introduction of the new energy source. Had *THEY* had access to ARV technology, there is no doubt that it would have been made public so that we could begin to counter the current global economic crisis. There simply is no sane rationale for continuing to allow further disintegration of the technologically developed world.

While it is said that it is impossible to prove a negative, the current economic meltdown comes about as close to proving that ARV does not exist as any argument could. The only alternative is to believe that *THEY* are prepared to allow national and global financial suicide, just to keep their secret. My ten trillion dollar challenge is for anyone to explain why that amount of money was willfully destroyed, if an answer was available.

As noted in my preelection prediction, ET did not get the message and the rescue ship will not arrive in time to save us from ourselves. Therefore, we had all better get to bailing as hard as we can and come up with solutions to the energy and economic crises that we have.

Other Considerations

If this conspiracy theory is to be believed, then the administrators must be myopic. As mentioned previously, over the past several decades the configurations of American aerospace industries have changed dramatically. Many of these changes occurred as defense contracts sharply declined. As these mergers and sell-offs of elements of the major companies occurred, the senior management understood they were in battles for institutional survival. Since it is most likely that ARV would have been constructed by a consortium of trusted companies, then the secrets of energy generation and propulsion would have been known in various parts of the industry. It is inconceivable that these companies, engaged in a life-and-death struggle for continued existence, would not have brought these capabilities forward as part of their diversification plans. If *THEY* had it and failed to transition the technologies to the civilian sector, *THEY* lost the opportunity to make trillions of dollars in revenues. When considering the total implications of withholding ARV technology from civilian applications, the lost income must be added to the dissolution of wealth previously mentioned.

As of the date of the writing of this book there is no plausible evidence that the hypothesized reverse engineering has been accomplished (or even attempted). The notion that the invisible *THEY* have accomplished the task would beg another critical issue—ownership! Even if these stories were true, then the reverse engineering most likely would have been derived by material from a crash that was recovered by some U.S. Government agency, probably the Army or Air Force. If so, the property belongs to the U.S. Government, and by fiat, the American people. That would include the rights to benefit from whatever discoveries were made from researching the UFO. If that energy source did exist it would comprise a gigantic opportunity for the country to solve many of the massive socioeconomic problems facing us, including both energy independence and financial stability.

While ARV might be rocket science, understanding this conspiracy theory is not. I will let common sense guide the reader as to the verisimilitude of the concept that *THEY* have successfully reverse engineered a UFO and are still withholding it from society.

Summary

Among the conspiracy theorists there are those who actually believe that some organization, be it the U.S. Government, private industry, or hybrid, has successfully reverse engineered a UFO. Called ARV, it is claimed that *THEY* are intentionally withholding these phenomenal technological advances that will be used later, most likely to enhance their wealth.

It is essential to understand that any such breakthrough would have implications far beyond small craft flying around the world. At the heart of the matter would be a new energy source. We know that the current geopolitical environment is dominated by dependence on petrochemicals. Despite efforts in development of green technologies, oil still fuels our economies. If ARV existed, it is inconceivable that *THEY* would have missed the opportunity to interject new energy technology into an economic system that allowed trillions of dollars to disappear.

In addition, to have such technology, conceal it from the public, and allow the military to sustain casualties needlessly would be treasonous. Very importantly, there is no credible evidence that ARV exists. However, if it did, it would belong to the American people as it would have been derived from public assets.

Although the story of ARV is not true, there are numerous UFO cases that are supported by strong evidence. The next chapter illuminates a few.

REAL CASES AND HARD DATA

The phenomenon of UFOs does exist, and it must be treated seriously.
—MIKHAIL GORBACHEV

There are so many strong UFO cases that selecting which ones to highlight is difficult. However, they are necessary to provide the reader with an appreciation of these high-strangeness and high-credibility incidents. Contrary to protestations from skeptics, there is physical evidence that corroborates the testimony of some very credible witnesses. Some of these cases are already well known; others will be seen for the first time. The strength will vary slightly. In a later chapter I will address issues associated with conflicting data that point to conclusions that are more complex than might be imagined.

Bentwaters

At the top of the list of important cases are the incidents that occurred in late December 1980 at the Royal Air Force Base at Bentwaters in the United Kingdom. Unlike many UFO cases, this is one that has gotten stronger over time. The quality of the firsthand witnesses is impeccable as most of them not only were highly qualified U.S. Air Force personnel, but were in the Personnel Reliability Program, better known as simply the PRP, as well. That means they had undergone extensive psychological testing prior to being assigned to sensitive positions.

For military personnel associated with nuclear weapon systems, being in the PRP involved both unscheduled physical testing (for drugs) and constant mental evaluations. One could be removed from the program for almost any reason, and therefore, participants became extremely risk averse. The near-universal perception was that reporting UFOs could cause supervisors to question one's mental stability—thus the loss of a job assignment. Also, in the U.S. Air Force, there was also the strong feeling that interactions with the Office of Special Investigations were to be avoided at all costs. The message was clear. Unless absolutely necessary, don't report sightings of high strangeness or that could not be explained. Still, in the Bentwaters case, some witnesses came forward immediately, and more have done so over time.

Very importantly, the case does not rely on a single fleeting incident but rather events that recurred over several nights and possibly longer than is commonly known. It also comes with substantial supporting physical evidence, including casts of indentations made in the frozen ground by the weight of the landed craft. There were radiation readings that were above normal taken at the site where the UFO damaged trees as it maneuvered away from the observers. There was also a recording made during the real-time investigation of one of the occurences. In addition, there was a short film made of an incident in the same area the following month. Unfortunately, there was too little detail or reference points to make much of it.

Touching the Craft—It Doesn't Get Any Closer

One of the most striking interviews conducted relative to the Bentwaters incident was when Staff Sergeant Jim Penniston responded to my question with, "You mean the one I touched?" That statement put a new perspective on the case. I had heard that the encounter was quite close, but physical contact with the craft was new and important data.

Jim Penniston was involved in the initial contact shortly after Christmas. He went into the field adjacent to the base as there were concerns that an aircraft might have crashed. As he entered into the woods he saw a luminescent object about seventy meters away with a white light of sufficient intensity to cause him to squint. It was unlike any aircraft he had seen before, and as a trained investigator Penniston made careful observations. The UFO was pyramidal in shape, approximately nine feet by nine feet by nine feet at the base and standing about six to seven feet in height. The exterior was black in color and to his touch felt warm. There seemed to be static electricity in the air. Surprisingly the covering was like smooth fabric and not metallic as is often portrayed. He noted the edges of the craft had colors of red, blue, and yellow, and that these lights would move about. On one side was glyphlike writing that he could not decipher. Penniston noted these symbols seemed to be part of the craft, not something engraved into the material. In his written notes taken at the site he indicated there was no apparent landing gear.

After he closely observed the object for an estimated fifteen minutes it levitated and began to move silently away from him through the trees. When it was about 200 feet back it lifted up through the trees and was gone in the blink of an eye. According to him there were about eighty witnesses to the departure of the craft.

Staff Sergeant Penniston was accompanied into the woods of Rendlesham Forest by John Burroughs and Edward Cabansag, both of whom wrote reports that are available. Like Penniston, Burroughs reported being physically close to the object while it was on the ground. The drawing he made is quite striking as it is congruent with the description made by Jim Penniston. Those who have heard Burroughs speak at UFO conferences are aware that this was an emotional event that continues to bother him.

Two reports written by their supervisors indicate the frame of mind of the witnesses. Lieutenant Fred Buran stated that Penniston "saw something out of the realm of explanation for him at that time." Master Sergeant J. D. Chandler, after talking with the witnesses, wrote, "I was sure they had observed something unusual."

Three nights later the command was having a Christmas party when the security patrol lieutenant busted in with the statement, "It's back." As the deputy base commander, Chuck Halt left the awards festivities to the commander and departed believing that he would make quick work of the matter. Little did he know that what was about to happen would change his life forever.

It is worth knowing that Chuck was an excellent officer and already on the fast track for promotion to full colonel. While UFO buffs seem to think this event helped his career, nothing could be further from the truth. In reality, after the infamous memo that he wrote the following day was released to the public, it nearly derailed him. He has stated that he believes this encounter had adverse impact on the military careers of some of the enlisted men involved. As has been pointed out throughout this book, in official circles encounters with UFOs are not considered a good thing, and never career enhancing.

Gathering some gear, including a recorder and an APN-27 Geiger counter to measure radiation, then Lieutenant Colonel Halt headed into the woods accompanied by Sergeant Nevels. As with the previous incident, the group could see a bright light in the forest. Chuck described the object as glowing like the sun but with a dark center. There was also something that appeared like dripping molten metal coming off the craft. As they approached the object moved away from them, navigating through the trees. Then it silently exploded, split into five separate pieces, and disappeared shooting off in different directions.

Shortly thereafter they noticed unusual lights maneuvering in the sky. Chuck noted they moved in a formation and appeared to be under intelligent control. The rapid turning movements, some at ninety degrees, exceeded any existing Air Force capability and the impact of high g-force turns on humans has already been covered. The lights were bright and multicolored. Suddenly they stopped and headed toward the group at high speed. Directly above them the objects again stopped. Then a bright laserlike beam was shined right at their feet for several seconds. Just as suddenly, the light went out, similar to a flashlight being turned off.

There were radiation readings taken that were higher than expected, but only at the site where the craft had been sitting. A UK MoD investigation would place the readings at least eight times above the normal background. Indentations of about three inches deep and with diameters of eight to ten inches were noted in the hardened ground. Plaster casts were made of each of the three points of contact. During the investigation Chuck also contacted the base to determine if their radar systems were picking up unidentified objects. They were not.

While in the field that night Chuck made a recording that has been released to the public. Much was made about the periodicity of the light with immediate speculation that the crew had mistaken the beam from a nearby lighthouse for the UFO. Chuck told me that they were well aware of the location of that lighthouse.

Years later, during a re-creation for a television program, skeptics accompanied them to the actual site of the encounter. When establishing ground truth, it was clear to everyone that they could not have seen the lighthouse from the actual location of the craft.

As some people know, the sightings did not end with the two events in December. Steve La Plume, with whom I spoke on the phone, also had a sighting on January 9, 1981. He described seeing aerial lights, again near the area called East Gate. The lights moved around in what he said were sinelike motions. That was followed by the spotting of a huge cylindrical-shaped craft that was estimated to be the size of an aircraft carrier.

Other people from the base were reportedly involved as well. While at NIDS contact was made by a woman who at the time of these incidents was married to one of the base's senior officers. As a matter of course they had access to the security police radio channels. That was not related to this incident but was done so that the officers could quickly hear of events developing on the base. She states that the sightings continued for some time, well into January 1981, and that some of the dependents got involved in following up on reports. As proof she provided a short film of one of the sightings.

There are other reports of UFOs moving over the storage area and sending down beams of light. In 1980, the joint U.S.–UK base at Bentwaters was home to nuclear weapons, a fact not divulged to the general public for reasons of security. Knowing about the weapons stored there, during my first interview with Chuck many years ago I asked if any warhead had been examined to determine if it had been tampered with in any way. My thinking was that these were strategic systems and it would be better to check than to ignore the potential intrusion. At that time, he said he did not believe a reliability check had been made. There are other reports suggesting that at a later time a C-141 Starlifter did come in and pick up some items. However, there is no indication of what might have prompted the operation. Such missions occurred from time to time but the base administration was not involved with any of the details.

Enter the Skeptics—Sans Data

Despite the strong evidence supporting this case, the response of skeptics demonstrates how even the best cases would be dealt with. Ian Ridpath was one of the major dissenters. Although he had not been on the ground that night, he quickly decided that it was the lighthouse that was seen. He also noted that the light coming down from the sky "was a bright fireball from space that was burning up in the atmosphere exactly that time. I think that is what they saw." He pronounced that the indentations that yielded plaster casts were rabbit holes, and the radiation readings were not mean significant. As Nick Pope, who went out and studied the case, later noted, "These must have been radioactive rabbits."

This approach exemplifies the "any answer resolves the issue" tactic that is used so often by skeptics. Never mind that the situations he described did not fit the facts. Remember, Penniston reports touching the craft. The timing of these events is not an exactitude, yet Ridpath knows the exact timing because of unrelated astronomical events. However, his hypothesis does not explain the light at their feet staying on for several seconds. It is known the lighthouse was not visible from the real landing site. Still too frequently self-appointed skeptics with nothing but a personal opinion get airtime just because they are willing to pontificate. Facts be damned.

The security implications of this case are significant. The primary participants varied about what they were told regarding classification. Both Penniston and Sergeant Nevels, who had been with Colonel Halt that night, indicated they were informed the subject was classified. Obviously Colonel Halt's memo was always unclassified and was released to the public. While there are some unconfirmed stories about unofficial observers being harshly interrogated, possibly with chemical agents, they seem unlikely. The difference in reports concerning classification of the incident is consistent with earlier comments I have made that low-level investigators simply wing it. When in doubt, call it classified. It is noteworthy that no one has been admonished for speaking publicly about their observations.

When Chuck retired from the Air Force he was routinely, but formally, debriefed on a number of classified programs on which he had knowledge. Bentwaters was not among them. When he asked if the intelligence personnel had any questions about the incident, or if there were prohibitions about speaking about it, he said, "They just laughed."

Cash-Landrum

Nearly half a world away, at about the same time as the important events at Bentwaters were breaking, another irrefutable UFO incident occurred. Not far from Houston, Texas, around 9:00 P.M. on December 29, 1980, Betty Cash and Vicky Landrum were driving south on Highway Farm Road 1485. In the backseat was Vicky's grandson, Colby Landrum, then seven years of age.

Initially the UFO appeared as a streak on the horizon. It came toward the car. According to John Schuessler's account, "With no warning at all, the sky seemed to split open and the object came angling down directly in front of them settling swiftly between the trees just ahead of them and above the highway." The object was described as a large rhomboid with "flames blasting down from the bottom." The light was brilliant and an intense heat could be felt. Uncharacteristic for most UFO encounters, a loud roaring sound was heard.

The driver, Betty Cash, immediately stopped the car less than 150 feet from the object and the heat continued to rise. Cash got out of the car and looked at the object. It is reported that Colby attempted to make a run to get away, but was caught by his grandmother and pushed back into the car. Vicky was out of the car, but partially shielded by standing behind the open door. At that distance the size was equated to a known water tower that measured 200 feet in height.

Being devout Christians, even at the time of the incident they regarded this as a religious experience. Cash was in front of the car and moved forward slightly. As she touched the fender she found it was "blistering hot." They stood there for an estimated three to five minutes. Then flames intensified and the object began to rise above the trees. As the UFO moved away, the trio became aware of approaching helicopters. They reported that the helicopters seemed to swarm around the object. They specifically identified some of them as CH-47 Chinooks, the kind flown by the U.S. Army and the Marine Corps.

However, Cash claimed they could read "United States Air Force" imprinted on them.

Irrefutable Physical Injuries

They went home, but that night they all began experiencing symptoms similar to radiation exposure. The degree of severity of the symptoms was directly proportional to the amount of physical exposure at the site, a fact that supports their story. Cash had the greatest degree as she had been in front of the car with her whole body presented to the flames. Later they suffered from nausea, vomiting, diarrhea, generalized weakness, a burning sensation in their eyes, and feeling as though they'd suffered sunburns. Within days, alopecia began.

For a more complete description of this event, John Schuessler's book, *The Cash-Landrum UFO Incident*, is recommended. John spent thirty-six years with the human spaceflight program, but also investigated many UFO cases. He spent considerable time and effort on this case.

The symptoms did not recede quickly. The victims required medical attention. The onset and severity of the reactions strongly indicated that they had been exposed to high-level radiation. The type of radiation raised new questions: Was it ionizing or not, and were infrared or ultraviolet frequencies included? Some researchers believe the victims' exposure was over LD 100, which means they should have died under normal circumstances.

There were other witnesses as well, but none as close to the UFO as Cash and the Landrums. Those other witnesses also confirmed the story of the helicopters, including the identification of Chinooks. It was this information that led to an investigation and a lawsuit against the U.S. Government and claims for twenty million dollars all combined.

The first inquiry went to the U.S. Air Force. However, since they do not have the Chinooks that were observed, the case came to the Army Inspector General's Office where it was assigned to a friend of mine, then Lieutenant Colonel George Sarran. To his credit, George did an excellent job with an investigation into a most unusual subject—one which had no prior guidelines. He took the case very seriously and began by tracking down all of the units that had the types of aircraft described.

The working hypothesis of the proponents that the U.S. Government was at fault was that the UFO must have been some form of experimental craft with a nuclear engine that had gotten out of control. The helicopters were there to shepherd the craft back to wherever it belonged.

There are several problems with that hypothesis. First was the timing. The incident took place between Christmas and New Year's. This is traditionally a period in which the military has stood down and no major exercises ever are scheduled. Next was the location. The government has places where they test experimental aircraft, including Area 51. They would never test a secret vehicle so close to a major metropolitan area.

George visited all the units that had similar helicopters. Fort Hood, Texas, has many helicopters but it is located several hundred miles away. He checked on Task Force 160, then a classified unit that flew special operations missions and was stationed at Fort Campbell, Kentucky. He even checked with the U.S. Marine Corps and found the amphibious fleet was in the Atlantic off Florida at the time of the incident. Being thorough, George made connections with the helicopter fleets of the oil companies that fly crews to the offshore rigs. The bottom line is that no helicopters could be located that could have been involved that evening.

George carried the investigation a step further asking for consultation from me, and two military medical doctors, U.S. Navy Captain Paul Tyler and Air Force Major Richard Niemtzow, both of whom specialized in radiation. Paul and I had worked together for several years in my interagency projects at INSCOM while Richard had prior experience with French UFO cases. Based on the physical evidence available, our conclusion was that the victims were telling the truth and had been exposed to high levels of radiation. However, this case simply defied any conventional explanation.

Mansfield, Ohio

Another famous example of a UFO interacting with a military helicopter occurred on October 18, 1973, near Mansfield, Ohio. This is one of the few cases that, like Bentwaters, got stronger as the investigation continued. The principal researcher was Jennie Zeidman and the details of the case were included in Peter Sturrock's book, *The UFO Enigma*. There were four Army Reserve personnel on board the HU1-H Huey with Captain Lawrence Coyne as pilot and Lieutenant Arrigo Jezzi as his copilot. At about 11:00 P.M. as they flew back from Columbus, toward Mansfield, Sergeant Robert Yanacsek spotted a red light that appeared to be paralleling their flight path.

The light began to close on the helicopter and Captain Coyne put the aircraft into a powered descent. Thinking they were on a collision course, Coyne increased the rate of descent and notified Mansfield control tower of the situation. Both the VHF and UHF radios then malfunctioned.

The light slowed and came to a hover over the helicopter. The crew members reported that they could see a large metallic cigar-shaped UFO that had a red light in front and a white light to the rear. There was also a green light emanating from the bottom and bathing the cockpit. Although the collective was in a maximum down position, the altimeter was registering a climb rate of 1,000 feet per minute. In layman's terms, the UFO was lifting the helicopter when it should have been dropping fast. This sounds much like a *Star Trek* tractor beam. The crew reported feeling a slight bump as the UFO released its hold. Importantly, they did not hear any noise or experience turbulence from the craft above them. Once released it was determined that they were well above their initial flying altitude when they should have been closer to the ground.

While the witnesses in the helicopter make a very strong case for this event, it turned out that there were civilian observers on the ground that could confirm the story. They saw the object approach the helicopter and confirmed that it seemed to lift it. Their physical description of the UFO and events fits quite well with that of the crew.

Though the incident lasted approximately five minutes, archskeptic Phil Klass stated the crew had misidentified a meteor crossing the sky. Of course meteors don't stop, hover, and lift helicopters. This is another example of a skeptic providing an answer that does not fit the facts, yet gets reported anyway. The helicopter interaction with a UFO over Mansfield, Ohio, still ranks as a bona fide high-strangeness/high-credibility case.

Eskimo Scouts

My first encounter with the Eskimo Scouts came in 1963 when attending Infantry Officer's Candidate School at Fort Benning, Georgia. Like most of the soldiers in the Army, I did not know they existed, nor had we heard of the 207th Infantry Group to which they belong. That initial experience was not great. More at home in a kayak hunting seals than in a classroom or rifle range, Sergeant Maniitok (a pseudonym) was far out of his element at what was jokingly called "the Benning School for Boys." To assist those who were struggling, the cadre assigned roommates, pairing the weak students with stronger ones. As I was first in the class, for the duration of the program I inherited Sergeant Maniitok, the weakest member of our entire group. To be kind I'll just say we operated on different value systems.

However, it was learned that the Eskimo Scouts were on continuous patrol in western Alaska. The troops were drawn from the Aleut, Athabascan, Inupiat, Haida, Tlingit, Tsimshian, and Yupik people and covered more than 200,000 square miles with a small force. Unbeknownst to most Americans, the Soviets would periodically probe Alaska. The country needed someone to be on guard and these natives were a perfect match. They lived just about a hundred miles from Siberia and came with a prestigious history of fighting in the Aleutian Islands during World War II. Even though they were a National Guard unit, these scouts were on duty 24/7. Wherever they went, whatever they were doing, it was their job to observe and report back if they saw anything out of line. As many of them survived by hunting and fishing, they were often in remote places—but rifles and radio were always handy. Of course, their very existence required extreme vigilance and keen powers of observation. They fully understood the environment and knew that a missed signal or abrupt weather change could mean the difference between life and death.

It is their powers of observation that make their reports so compelling. In the late-1980s a friend of mine in the Inspector General's Office conducted a visit to Alaska. While at a battalion command center he ran across a stack of UFO reports that had accumulated there. They are significant for two reasons. One is the content of the reports. The second is that they were taken and then forwarded to that headquarters where they just sat accumulating dust. The point is, there was no place for the battalion to send the reports.

There were eighteen reports of events in January 1987 near St. Lawrence Island, which lies in the Bering Sea well west of the Alaskan mainland and adjacent to the international date line. This rather large island is home to just more than a thousand people, mostly Yupik, who are among those who hunt for survival purposes. The following are examples of their reports from that time frame.

> At picture A is a replica of the drawing submitted. This took place at 5:45 P.M. on 28 January and is one of the best drawings made. The large domed object on the right side of the picture was described as far larger than a Boeing 747 aircraft. It was followed by two smaller discs. All were very dark on the underside and were traveling toward the north trailing smoke. The observers then noticed that the large craft appeared to generate a cloud that engulfed it. The UFO, shrouded in the cloud was then seen to fly against direction from which the wind was blowing.
>
> As depicted in drawing B, on 29 January at about 7:30 P.M. observers on the ground spotted an airplanelike object flying from west to east near the tiny town of Gambell on the northwest corner of the island. This was described as having bright white lights at the nose and two in the tail area. In the center area of the UFO were green and yellow lights. This was unlike any aircraft they had seen before.

Drawing C shows a craft similar to B, and was observed at about 9:00 P.M. This was also moving west to east and was seen at Savounga, the only other town on the island. This picture shows the body of the UFO being thicker than the prior one.

While the intelligence officers at battalion headquarters could not identify the craft, they entered an interesting observation. The timing of these incidents was prior to the declassification of the F-117 stealth fighter. However, as mentioned earlier, there was considerable speculation about what the craft might look like and the toy model F-19 had just hit the market. Falling for the cover story, the officers believed that what was reported was a stealth fighter. It wasn't. There are a few other sightings as well, but my main point was while reports sometimes come in, there is no formal reporting channel in which to enter them.

The Phoenix Lights

The name given to this event is both a misnomer and misleading. The official response to this event, or series of events, is as important as the observations. First, while thousands of people in Phoenix did see the UFO, so did people in many other areas. In addition, there is strong evidence that this was not a singular event though the best known occurred on March 13, 1997.

Long-distance Event

The first report came at about 7 P.M. from Henderson, Nevada, which is contiguous to Las Vegas. Following that there were sightings of a gigantic, slow-moving craft that were time-synchronized over Kingman, Arizona, southeast to Prescott, on to Phoenix, and then as far south as Tucson. There were videos and photographs taken as the object slipped silently along. Importantly, witnesses noted that as the object passed overhead, the stars could not be seen. That means the UFO was a solid object and could not possibly be light aircraft flying in formation as some have alleged.

In the Phoenix area there were at least two sets of sightings that night, and maybe more. The key sightings took place between 8 and 9 P.M., and no convincing explanation has ever been made. When questioned, officials at Luke Air Force Base, which is located immediately next to Phoenix, put out a news release attributing the incident to flares dropped from A-10 Warthogs of the Maryland Air National Guard, which was practicing over Barry Goldwater Range just to the southwest of the city. Unfortunately it took them four months to announce the flare theory. Importantly, the timing was off as the flares were dropped after 10 P.M. Of course skeptics quickly glommed onto that explanation and pronounced the case solved. Again, the facts did not fit the circumstances in time or location. Certainly flares did not drift the distance of more than 300 miles that the V-shaped UFO had traveled nor would they block out the stars.

Unconscionably, in June 1997 then Governor of Arizona Fife Symington held a press conference at which he had an aide enter dressed in a space alien costume. It was clear at that time that he was belittling the whole experience by making fun of a situation that no one understood. A decade later, Symington made a public admission; he had been one of those witnesses that night in 1997. At about 8:30 he too had seen a large V-shaped craft that blocked out the stars just as the others had reported and was now willing to discuss the matter on national television, including *Larry King Live*. His excuse for the 1997 press charade was concern about how emotionally witnesses were reacting to the event. He was bothered by the fact that he could not get any good answers from the U.S. Air Force, his own Adjutant General, or the Arizona Director of Public Safety.

Despite his own later legal problems, Symington made a pretty good witness. Of his own sighting he stated, "As a pilot and a former Air Force officer, I can definitively say that this craft did not resemble any man-made object I'd ever seen."

For a complete description of this intriguing case I recommend either the award-winning documentary *The Phoenix Lights: We Are Not Alone* by Dr. Lynne Kitei, or her recent book, *The Phoenix Lights: A Skeptic's Discovery That We Are Not Alone*. As she lives in the Phoenix area, Lynne has been able to record and study many similar luminescent phenomena over a period of years. In fact, she has photos in which orbs of light are seen as she is looking down from her home in the hills nearby. See her Web site at www.thephoenixlights.net for more details.

Gulf Breeze

The incidents at Gulf Breeze, Florida, make a fascinating study. The case came to light in November 1987 when photos taken by Ed Walters were published in a local newspaper. While much has been made of those photographs, along with claims of fraud, there remains a core story that seems to elude most researchers. As the story goes, Walters had given his photos to the editor who had decided to run with the article. Gulf Breeze is a very small town near the western end of the Florida Panhandle so the national impact of any story would be minimal. As the editor prepared the edition, he was visited by his parents.

They saw the photos and inquired where he had acquired them. They then indicated to their son, "That looks like the thing we saw." For the editor, that sealed it. If he couldn't trust his own parents, then who could be trusted?

For Walters, this was not a singular event. Rather, he continued to have sightings and take more pictures. That alone was troubling for most UFO researchers. The feeling was, once maybe. When contact continues, then the researchers become very suspicious and generally reject the case altogether. Among the researchers to get involved with the photo analysis was Dr. Bruce Maccabee, then an optical physicist with the U.S. Navy. Bruce examined the pictures and did not find any evidence that they had been faked. There were some aspects of the photos that did seem unusual.

But, as Bruce notes, this is not just a photographic case. There is much more evidence than relying on Walters's sightings. As the story broke, more and more people came forward. A little over a year later Maccabee wrote a paper entitled "Gulf Breeze Without Ed." By that time there were 55 additional witnesses and many of them reported seeing the craft when Walters was not around. In fact, sightings were sufficiently common that people began a nightly vigil at the shoreline. Eventually more than two hundred independent witnesses would come forward. As Bruce once mentioned to me, what does it take to cause people to change their lifestyle to accommodate UFO sightings? That happened in Gulf Breeze.

This case is unique in that sightings occurred periodically for several years. The local people even formed a research team that according to Maccabee, "logged about 170 sightings, most of which involved multiple witnesses and most of which included still photography with telephoto lenses and/or recording by video cameras. And, in several cases a light was observed simultaneously by two separated groups of people thereby allowing for triangulation."

There is a *Men in Black* (MIB) anecdote that goes with this case. In the early 1990s, while working at Los Alamos, I was involved in a project with the Lockheed Skunk Works. This entailed traveling with three of their engineers to the U.S. Special Operations Command Headquarters at MacDill Air Force Base in Tampa to give a briefing. The following day we were to travel to Hurlburt Field, which was located about 50 miles east of Gulf Breeze. In making the travel arrangements, hotel reservations were intentionally set near Gulf Breeze. The schedule was that we would fly from Tampa to Pensacola late in the afternoon and drive to Hurlburt Field in the morning. The intent was to drop the Skunk Works men at the hotel, then slip out to the beach by myself.

Unfortunately, late-afternoon thunderstorms hit Tampa just as we were about to leave. Thus we were delayed several hours and did not arrive in Pensacola until shortly after dark. Having come directly from the briefing at MacDill, we were still all dressed in dark suits and ties. To make matters more interesting, the rental car company gave me a vehicle that was such a dark color of green it was indistinguishable from black. Everyone knows that MIBs come in black sedans.

My friends agreed to accompany me to the beach and professed modest interest in the topic. So there we were, dark suits from LANL and the Skunk Works, in a black sedan. What could be more fitting for the local lore? To say the local audience was skeptical of our intent would be an understatement.

Still, a lot was learned about the phenomena of Gulf Breeze that night. Many of the people on the beach opened up and told us about what transpires. They told us enthusiastically about a sighting that Bruce Maccabee had at that site. They also described a wide variety of occurrences, far more than were generally associated with the photos. Among the most interesting were stories of balls of light that came down very close to them and would at times pass between people.

Ed Walters and the photos aside, the Gulf Breeze case is important for what it did for the community. Years after the initial reports, here were people who were altering their lives so that they could come to the beach on a regular basis. Whatever the phenomena was, it was able to captivate their attention and hold it for a long time. That is an important and untold aspect of the Gulf Breeze sightings. A detailed discussion of the case can be found on Bruce's Web site, *www.brumac .8k.com/GulfBreeze/Bubba/GBBUBBA.html*.

Strategic Missile Systems

There were documented interactions between UFOs and missiles both in America and the Union of Soviet Socialists Republic. Many of those incidents were simple sightings of UFOs near the missile. Undoubtedly some of the sightings were misidentifications of known effects associated with launches. However, there are two cases that stand out in the annals of UFO history that defy any prosaic explanation and are of undeniable strategic defense implications.

Malmstrom Air Force Base

The U.S. incident took place at missile sites under the command of Malmstrom Air Force Base, Montana, on March 16, 1967. Since then, Robert Salas, who was actually in one of the underground command centers, has written about this incident and has spoken at several conferences. He is a highly credible witness. In addition, I had the opportunity to interview many of the other participants. These were highly qualified personnel, some were launch control officers, and all were in the PRP. Among those I talked with were the men who came to repair the site, contract engineers investigating the incident, and a person who was aboveground and confirms the relationship between the missile failures and the UFO sighting.

The strategic significance is that our intercontinental ballistic missiles (ICBMs) comprised one-third of the nuclear triad. That triad consisted of the Ballistic Missile Submarine fleet (SSBNs), the manned bombers (B-52s), and these underground missile bases. These units were our deterrence against attacks by the USSR and nothing short of national survival was considered to be at stake. Malmstrom was part of the Strategic Air Command (SAC) which was responsible for two legs of the triad.

Located in central Montana, two complete flights, Echo and Oscar, were involved in this incident. Each flight controlled ten ICBMs, an awesome responsibility. The first incident occurred at the Echo Flight site. At about eight-thirty in the morning one by one each of the missiles went offline, meaning they could not have been launched. All ten went down in short order. That alone was unheard of. While missiles did periodically experience malfunctions and would shut down, that usually happened to only one missile in the flight at a time. All equipment systems have failures. They are known and measured by a procedure known as mean time between failure or simply MTBF. The anticipated failure rate for the ICBMs was one missile in the entire squadron every six months. What happened at Echo Flight was off the charts in probability. In this case the failure was indicated as a fault in the guidance and control system, yet power had not been lost.

The Security Team Reported a UFO Near the Silos

Deep in the bunker, the Deputy Crew Commander stated he then received a call from the security personnel who were stationed aboveground. The guard reported that a UFO had been hovering over the launch site. Of course the guard was not aware of the situation below. The report was received by the officers with considerable skepticism. Two security alert teams also reported that the maintenance personnel all had observed UFOs hovering over two of the sites.

That same morning a similar incident happened at Oscar Flight where Bob Salas was situated in the underground Launch Control Center. Before dawn the Airmen aboveground saw UFOs maneuvering in the sky. The objects, which were not known aircraft, reportedly would streak across the sky, stop, change directions, then return to a position overhead. This information was relayed down to Bob, but he did not think much about it.

Shortly thereafter Bob received another call. This time the security NCO reported that a UFO was hovering by the front gate, and requested instructions. Waking his colleague, Lieutenant Fred Meiwald, Bob began to brief him on the calls when the klaxon went off, alerting them to problems with their missiles. They no longer had time to deal with the UFOs that were frightening the guards. They reported that within a few seconds they had six to eight of the ten Oscar Flight missiles go offline. Note that Meiwald, whom I also interviewed, confirmed the details of the incident.

At this point the squadron had at least sixteen missiles in a no-go configuration; something that has never happened before or since. It was stated that SAC Headquarters was frantic at hearing of these events. There is no simple reset switch that would bring the missiles back into operation. Among the people interviewed was the commander of the following shift who spent an entire day resolving the problems.

There was an extensive investigation led by a Boeing technical representative, with whom I also spoke. According to him, they could find nothing in the system that had caused the failures. Importantly, there were no power outages that accounted for any of the actions. The best guess was that some form of electromagnetic pulse, or EMP, had been employed. The problem with that suggestion was the size of the equipment that would have been required, and it would have to have been placed very close to the silos. The proximity of security and maintenance personnel would seem to rule that out.

The good news is that such an event never happened again. Of course the bad news was that the cause was never determined. For more detailed information about this case, articles by Robert Salas are recommended.

Byelokoroviche/Khmelitskiy

George Knapp was the first reporter to learn that the United States was not alone in experiencing problems with UFO-related interference with strategic missile systems. In fact, when he first heard of the incident, it was still classified by the Soviets and the people of the USSR had not been informed about it. He had gained access and the confidence of senior Russian officers who confided in him about this bizarre incident and alluded to several others. As in the Malmstrom incident, there is little doubt about a connection between UFOs and the interaction with the missile system. It is worth noting that the Soviet approach to the UFO issue had both differences and similarities with that of the United States. On directions that came initially from

Joseph Stalin, they actively sought observations from military personnel and civilians alike. However, there were intermediaries in the Kremlin who were quite skeptical of the whole idea and actively promoted information that UFO stories were a ruse of the West. As in the United States, they both ridiculed the scientists that were involved in the studies, yet covertly asked them to provide briefings.

George was fortunate enough to obtain in-depth interviews with Colonel Boris Sokolov who headed UFO investigations for ten years. Sokolov told George that they had more than six million cases on file. Only a very small number of them were considered high-strangeness/high-credibility. As in America, they found about 5 percent of the reports to be truly unidentified. In later years, by encouraging people to report sightings voluntarily, the Soviets established a massive network that made observations of their vast territory. Importantly, the cost of the operation was very low. It was well after Knapp's report was made public that ABC News took his story without attribution and went to the area to follow up. There they conducted interviews with many witnesses and the segment aired on the network show *Primetime* in October 1995.

The incident began at about 6:00 P.M. on October 4, 1982. The huge UFO was seen by hundreds of people as it hovered for hours near a Soviet missile base at Byelokoroviche. (Some researchers have identified the site as Khmelitskiy, but it is in the same location in Ukraine.) The unidentified craft was described as being large; 900 feet in diameter, possibly larger by some accounts. One key witness, Lieutenant Colonel Vladimir Plantonev, described the UFO as being similar to what was depicted in the movies, but perfectly smooth and without any apparent openings.

Lieutenant Colonel Vladimir Plantonev was on duty during that period of time an unknown force took over the launch operations and began to prepare the missiles for a strike against the United States. According to his account, for about fifteen seconds their monitors indicated that the missiles were winding up. A centralized system, this should have been controlled from the Kremlin, but no such orders had been given nor had the launch codes been entered into the computer. This situation terrified the launch control officers as they were completely unable to stop the missiles from preparing to launch. Based on the mutually assured destruction (MAD) concept in effect at the time, they were correctly gravely concerned that the United States would retaliate massively, believing the Soviets had initiated a nuclear war via a surprise attack. Then, as suddenly as it began, the system shut itself down, also without human intervention.

Investigators from Moscow under the direction of Colonel Sokolov totally disassembled the site but failed to find a rational explanation—such as an electronic malfunction—for the event. Colonel Igor Chernovsky, who accompanied Sokolov on the investigation, indicated that the actions should have required an intervention from Moscow to allow this incident to have happened. That the Kremlin's control had been usurped was extremely troubling to them.

American skeptics noted that some flares were reported by some of the witnesses. In fact, the Russian military had dropped flares, but that was totally unrelated to the missile incident. In fact, all of the formal statements concerning the event were made by Russian military officers. They obviously knew the difference between a 900-foot UFO overhead and flares dropped many miles away. The skeptics, who never visited the site, concluded that it was maintenance failure.

That does not address the large number of people who saw the large UFO hovering over the site. As for Colonel Sokolov, on camera he told George that they officially had no explanation for the event. The unofficial comment George states was "that the UFO was sending a message, making it clear that someone else could control our most powerful weapons systems and that we were powerless to do anything about it." George notes that they were pretty shook up. He also put the sensitivity of the revelation of the event into perspective. Had Colonel Sokolov provided such material to Knapp five years earlier, he would have been jailed; while ten years prior, he would likely have been shot.

High Over Dallas (and Big)

There are many high-quality cases involving observations by airline pilots. Dr. Richard Haines, a retired scientist at NASA–Ames, has collected thousands of them and his studies are highly recommended. Here is one case that serves as an excellent example of a high-strangeness/high-credibility incident. Fortunately it had multiple witnesses in two commercial aircraft and they were willing to talk about what they saw. It happened while I was still with NIDS and I had the opportunity to personally interview the pilots from both planes.

The incident took place on October 26, 1999, at about 2:30 A.M. CST in the air corridor high over Dallas–Fort Worth, Texas. Captain Karl Lenker was in command of a Delta Airlines flight headed east at 37,000 feet. At the same time Jose Rodriguez was the copilot on a United Airlines flight also headed east at about the same altitude. Both pilots reported seeing a large object that was brightly lit to their ten o'clock position. At first, Lenker believed that there were three separate objects flying in a V formation at a distance of approximately five to ten miles and heading in the same direction as the commercial airliners. He stated it was the brightness of the lights at this altitude that attracted his attention to them.

Lenker contacted the FAA and asked if they had contact with anything in that area. They confirmed that they had no such aircraft in the area. At this altitude the flights were in airspace designated for commercial use. The FAA also indicated they did not have any military aircraft operating in their vicinity. After several minutes the UFO made a slow turn to the north. Lenker noted that the geometry of the turn made it appear to be a single craft, although he could not see any connecting structure.

Rodriguez, in the United flight, was flying southwest of Lenker. Thus he could see both the Delta flight and the UFO. Lenker, however, could not see the United flight. Rodriguez reported that the aircraft was very large, "much larger than an Air Force C-5," an aircraft with which he was familiar. He also stated that as the lights dimmed when the UFO turned to the north, he could see a connecting structure among the lights. That means the object was a single large craft, not three UFOs traveling in close formation.

According to the geometry, the UFO was closer to the Delta flight than the United airplane. Rodriguez stated that he held up his hand to get a rough measurement of the size of the UFO. It was about as wide as his fist at arm's length. Given the known distance between the aircraft, but not the distance from the UFO, rough triangulation still can be done. The most conservative estimate makes the UFO more than a mile across. That is a very big object to be flying unannounced in commercial airspace.

Also, both pilots were former U.S. Air Force officers and had extensive experience with night flying and air-to-air refueling techniques. They noted that the Air Force does illuminate during refueling exercises, but they are normally done at much lower altitudes, and never in the commercial lanes. Both pilots indicated to the FAA that they wanted to make a formal report, something that rarely occurs. Unfortunately, at that time there was no official government reporting channel available.

Soviet Airspace

"Skeptics and believers both can take this as official confirmation of the existence of UFOs." This statement, made in March 1990 by Colonel General Igor Maltsev, Chief of Staff of the Soviet Air Defense Forces, is astonishing to say the

least. After all, his predecessor and Defense Minister Sergei Sokolov were very publicly fired for a relatively minor intrusion just three years prior. In May 1987, a German pilot, Mathias Rust, who was not yet 19 years of age, flew though Soviet airspace, avoiding detection and landing his Cessna-172 on the bridge at Saint Basil's Cathedral near Red Square in Moscow. That caused an uproar, yet UFOs flying about were somehow acceptable. Well, maybe more like something they could not do anything about. This position was confirmed by General Ivan Tretyak, who was then the Soviet Deputy Minister of Defense as well as Commander in Chief of the Air Defense Forces.

UFOs and Airplanes Get Very Different Treatment

General Tretyak took a page from the U.S. Air Force position and stated that UFOs did not appear to pose a threat, although their origin was unknown. This is a surprising statement as the Soviet Air Force had shot down several intruders over the years. Those attacks included the infamous case on September 1, 1983, in which Korean Airline Flight 007 was destroyed over the Sea of Japan killing all 269 civilians aboard. When asked why he had not given an order to fire on the UFOs, he stated, "It would be foolhardy to launch an unprovoked attack against an object that may possess formidable capabilities for retaliation."

He also stated that UFOs had been photographed by interceptor pilots and confirmed on both optical and thermal sensors but sometimes appeared to have stealthlike capabilities to evade radar. Some of the incidents occurred for longer periods of time and General Tretyak described an encounter that lasted three and a half hours. Other reports indicate that Soviet pilots actually flew over UFOs. Both Tretyak and Maltsev took care to note that the vast majority of cases are either misidentification of natural phenomena or hoaxes. When pressed about his interest in UFO reports, Tretyak stated it raised "moderate curiosity." Echoing the comments of his Chief of Staff, Maltsev, he stated, *"There are real phenomena of some kind which are appearing before us in the form of UFOs, the nature of which we do not know."*

Like American observations, the Soviet Air Force had many instances in which UFOs were observed hovering, then departing at great speed. Some of these reports estimated the craft to be 100 to 200 meters in diameter (more than 600 feet) with speeds ranging from hovering to triple the capability of modern fighters, yet then stopping again instantaneously. Other reports indicated that the UFOs displayed "startling maneuverability," yet made no sound. An investigation claimed that the UFO observed "was completely devoid of inertia. In other words, they had somehow 'come to terms' with gravity. At the present time, terrestrial machines can hardly have such capabilities." This means the Soviet scientist believed that the UFOs had perfect antigravity, which would represent a major technological breakthrough—one that still eludes scientists on Earth.

Like the United States, the Soviets also allowed UFO stories to cover classified projects. These included UFO reports when missiles were being launched. Some of these incidents were known to the ATP study group. For the record, the translation of these Russian reports was made by FBIS, the Foreign Broadcast Information Service of the CIA.

Blue Book in Default

Contrary to the opinion of most supporters of the Condon Report, Project Blue Book was not run by a scientific body that conducted in-depth investigations of the most perplexing cases. Rather, it was a very small group with one scientific advisor, Allen Hynek, and functioned primarily in a public relations capacity. Reports arrived—or didn't—based on the personal opinion of local officers assigned to the UFO duties as an additional duty. All the local officers had an unrelated primary job, and probably several other "additional duties." In general, such additional duties are viewed as distractions from their main assignment. In short, being assigned to UFO reporting was not a good thing.

We know that many good cases did not show up in the Blue Book files. Conspiracy theorists see this as a control group extracting cases they deemed important. I suspect that most of those incidents were neglected, as this was not a high-priority function of the understaffed Project Blue Book office at Wright-Patterson Air Force Base. This is a short version of a case in point.

On October 24, 1968, there was a remarkable event involving at least sixteen highly credible witnesses, radar confirmation of an object tracking a plane, and a UFO landing near a missile base in Minot, North Dakota. Again most of the witnesses were in the PRP as nuclear weapons systems were involved. The incident began with a missile repairman observing a bright orange red light that was on the ground. He saw an object lift off and maintain a position parallel to his vehicle movement. Alarmed, he called back to the command center at the base.

There was a B-52 bomber involved in practice maneuvers in the general area. The control tower contacted it and gave the crew a vector toward the area where the UFO had been sighted. When the crew inquired about what they were looking for they were told, "You'll know it if you see it." The copilot at the time was Captain Brad Runyon, who would later come forward with this amazing story. Also on board was the radar officer, Captain Patrick

McCaslin. Following instructions he turned on a camera that would record the incident as it played out on the radar screen. Both the ground radar and that on the B-52 noted that there was an unidentified object following the aircraft.

The initial radar returns calculated the size to be similar to a KC-135 tanker, and the speed of the UFO at approximately 3,000 mph as it approached from the right to the left. The object then maintained a relatively constant position in the vicinity of the left wing of the bomber. It was reported that when the UFO came close to the plane, neither of the two transmitters would function properly.

Both the maintenance man and the B-52 crew confirmed seeing an object that was brightly illuminated. That ground observer estimated the object was about a 1,000-feet high, while the crew believed it to be either hovering near the ground or that it had landed. According to the Blue Book report an additional fourteen witnesses on the ground saw the light. At one of the missile sites an alarm was sounded. Investigation revealed that there was an open outer door and that a combination lock on the inner door had been moved.

While there was considerable detail noted during the incident, Lieutenant Colonel Hector Quintanilla, Director of Project Blue Book, appears to have ignored the facts. His conclusion was, "The ground visual sighting appears to be of the star Sirius and the B-52 which was flying in the area. The B-52 radar contact and temporary loss of UHF transmission could be attributed to plasma, similar to ball lightning. The air-visual from the B-52 could be the star Vega which was on the horizon at the time, or it could be a light on the ground, or possible plasma."

The B-52 Crew Describes Flying Over a UFO Resting on the Ground

From what Brad Runyon stated years later, he would agree with lights on the ground, but not mistaking car headlights as Quintanilla meant. In a television interview Runyon described a bright metal object 200 feet in diameter and several hundred feet long. There was a cylinder attached to a section that resembled a crescent moon. There was a yellow glow to the UFO and he then stated that he was "*fairly sure I was looking at an alien spaceship from another planet.*"

Runyon has added some details since his original report. However, it should be very clear that he was not mistaking Vega on the horizon for this massive craft below them on the ground. The original reports from this well-documented case can be found on the Internet. The incredulous evaluation by Quintanilla does provide some understanding of why aviators might have been reluctant to file reports to his office.

Watching War

This previously unreported case just recently came to light and involves UFOs observing military operations during the Vietnam War. During April 1970 then Master Sergeant Lou Rothenstein was an observer and photographer flying special operations intelligence missions in the backseat of an O-1 Bird Dog, a light observation airplane. At about 9:00 P.M. one evening, Lou and the pilot were flying east along the Vinh Te Canal, which runs close to the Cambodian border in the Mekong Delta. On this occasion they became aware of a mysterious, darkly painted French-built Alouette helicopter that was tailing them. Who was piloting the aircraft and how it managed to slip in behind them from Cambodia was never determined.

But that was not a UFO. As they proceeded toward the Seven Mountains region there suddenly appeared two real UFOs. When Lou initially spotted them, one was extremely close and at an estimated 200 feet from the tiny Bird Dog. Basing his guess on a known Army aircraft, Lou thought the nearest UFO was as large as a military Caribou, which has a wingspan of ninety-five feet. That was dangerously close for aircraft that do not have communications with each other. While the UFOs were a dark metallic color, there did appear to be illumination around the periphery. The Bird Dog was flying at about 2,500 feet altitude and the UFOs were about the same level and thus could be seen clearly against the backdrop of the taller mountains to the southeast.

Lou had the distinct impression that "the UFOs were looking at them" and observing their operation. The pilot in the trailing Alouette seemed to also spot the UFOs and quickly broke off his mission, diving and heading north across the border toward the Cambodian capital, Phnom Penh. At first the UFOs moved slowly toward the east and then rapidly accelerated up into the night sky, disappearing at a rate that was faster than any jet fighter.

As a retired command sergeant major, Lou is willing to describe the incident. He also stated that he and the pilot had a discussion about whether or not to report their sightings at the time. The decision was to report the unknown Alouette helicopter, but not to mention a word about the UFOs. As is often the case, they were concerned about their personal credibility. He remains convinced that there were many other UFO sightings that were never reported for the same reason.

Other Cases

Choosing which cases to include was very difficult as there are many from which the selection could be made. For most of the cases presented I either interviewed witnesses myself, or was in contact with the investigators. The reality is that UFO sightings occur globally. Many other countries have solid cases as well.

Brazil

Brazil has had its share of UFO cases. One sighting that came during the ATP investigation took place on May 19, 1986, and involved fighter interceptors being launched from multiple bases. There were both aerial and ground observations and radar contacts over a period of more than four hours. We were intrigued when the Minister of Aeronautics made public statements confirming the events. He stated that, "at least twenty objects were detected by Brazilian radars. They saturated the radars and interrupted traffic in the area. Radar doesn't have optical illusions. We can only give technical explanations and we don't have them."

In 2008 a most surprising encounter occurred in Paris involving Brazilian friends who own a nonlethal weapons company. While attending a huge arms conference called Eurosatory, the discussion of UFOs was frequently a focus of our hourlong cross-city commute. Fairly new to the topic was Vice Admiral Pierantoni Gamboa, who had retired from the Brazilian navy. One morning the Brazilian Ambassador to France, Jose Mauricio Bustani, came by the booth and the admiral signaled that he wanted to make an introduction. However, instead of mentioning nonlethal weapons, he told the ambassador about my interest in UFOs. To everyone's surprise the ambassador said, "Oh yes, I've seen them three times." He went on to openly describe what he had observed near his home in western Brazil. That is the only time that I've ever heard a politician speak so freely about the topic in such a public session.

Many other cases have been captured by A. J. Gevaerd and published in his magazine. From my discussions with Brazilians during my business trips there, I can say there is no shortage of interest in the topic, and many cases that are rarely heard of outside of that emerging economic giant. In attempting to appear more transparent, in August 2010 the Brazilian Government issued orders to their Air Force that all UFO sightings were to be documented. However, it was also indicated that they would not be chasing UFOs, or even investigating the reports.

China

China too has numerous reports of UFOs, as well as historical interest in mysterious objects. One case on October 19, 1998, involved a cat and mouse game played between a Chinese Air Force Jianjiao-6 interceptor and a UFO that had appeared on four ground radar systems. Supported by more than 100 witnesses on the ground, the pilots reported a mushroomlike dome on top and rotating brightly colored lights along the flat bottom. The UFO won. Most interesting is that it was the Chinese Government that made the announcement of the incident.

Chile

Taking a lead, Chile came forward more than a decade ago and declared that they would actively study UFO cases. The Chilean Air Force announced the formation of CEFAA, the name of which translates into English as the Committee for the Study of Anomalous Aerial Phenomena. CEFAA is attached to the Chilean equivalent of the FAA. This came about as several senior officials in Chile had personal observations of OVNI as they are called in Spanish. One noteworthy case was reported in 1988 when a Boeing 737 was forced to divert from a landing at Puerto Montt. While on final approach, the pilot suddenly was confronted with a large luminous object. This again points to aviation safety issues.

Peru

To the north, Peru also has experienced many sightings. There is an extraordinary case that is now public. When we first heard of this incident, the information was still classified. On April 11, 1980, Peruvian Air Force pilot Oscar Santa Maria Huertas was flying a Sukhoi-22 fighter when he was ordered to engage a UFO initially hovering in restricted airspace. The object was at first thought to be a balloon, but it proved to be capable of maneuvering away from him at supersonic speeds. He attempted to get above the UFO but was unable to do so, even at an altitude of 63,000 feet.

This report got our attention as it was one of the few in which shots were fired. Huertas states that he fired sixty-four 30 mm rounds at the UFO. Although many of the rounds hit the object, it sustained no visible damage. The bullets did not bounce off of the covering but rather it seemed to absorb them. Huertas described the UFO as metal, circular, 30 feet in diameter, and devoid of typical aviation fixtures. The craft remained visible for more than two hours and, according to Huertas, "This object performed maneuvers that defied the laws of aerodynamics."

Turkey

Turkey has been the site of numerous sightings, some of which have been captured on film. Turkish Air Force pilots reported being intercepted by light sources while on a training mission over the Aegean Sea. Haktan Akdogan has established a research center in Istanbul that takes these events quite seriously.

Mexico

Mexico has experienced many UFO sightings that demonstrate that concerns about aviation safety should be taken seriously. On July 28, 1994, Captain Cervantes Ruano was bringing Aeromexico Flight 129 in for a landing in Mexico City. As the plane descended through 5,000 feet and lowered the landing gear, it struck a hard metal object. Fortunately the plane landed safely. Less than two weeks later, on August 8, another civilian airliner unexpectedly encountered a metal disc that was more than 45 feet in diameter and hovering near the flight path at about 12,000 feet. Jaime Maussan, a well-known Mexican investigative reporter, has located many authentic cases. One of the most important was confirmed by Minister of Defense General Clemente Vegas Garcia. This involved an encounter with multiple UFOs that were recorded on both the airborne radar and their FLIR system. It was on April 20, 2004, that this counternarcotics plane, equipped with advanced sensors, made the sighting.

Germany

Germany is the focal point for MUFON-CES or Central European Section. Headed by Illobrand von Ludwiger, they have conducted many investigations of UFOs. As an example, on September 6, 1979, three objects believed to be about 45 feet in diameter were observed near Ingolstadt, Germany. Witnesses included police officers from several jurisdictions and many civilians. Photos were taken, and every object was described as having five points, each with a flashing red light and having a silver foil-like exterior. The Munich air traffic control notified a private civilian airplane transiting the area en route to Brussels. The pilot reported that four or five lights came swiftly at her as she flew at an altitude of about 11,000 feet. She reported that the lights circled her aircraft, then after about half a minute in close proximity, they simply disappeared.

Antarctica

Though they are fairly rare there are even sightings in Antarctica. U.S. Navy Captain Paul Tyler, who was mentioned in the Cash-Landrum case, has spent five tours wintering over on that frozen continent that has no indigenous population. As a member of the ATP project, Paul told us about observations made on the all-sky cameras that photographed the heavens for long periods.

He reported that periodically they would find tracks of objects that were not supposed to be there. There have been recent sightings at the research bases there as well.

Talk to Friends

Sometimes surprising cases come to light during casual conversations. The following one was told to me at a small party when I mentioned an interest in UFOs. It is offered as an encouragement for readers to discuss these matters with their friends. Wayne spent his entire adult career with the U.S. State Department and was a very senior officer when he retired. Based on his background I took him to be a highly reliable witness.

Wayne reported that when about 9 years of age, he and a group of children were being driven to a place west of Albuquerque on the old Route 66. There were multiple cars and a total of about 15 people in the group. Sitting in the back of the car he noted that a large UFO was hovering above them as they continued to the west. He described the object as circular and about 50 feet in diameter. It was somewhat nerve-racking as it was at a relatively close range.

The drivers of all of the cars pulled off the road and they watched the UFO for about 25 to 30 minutes. The craft went to the south side of the road and landed at a lower elevation. Landing was accomplished on three limbs that were extended as the UFO approached the ground. They could see human-like figures moving about inside the craft through the round portholes. No one exited the craft and eventually it rose up and then accelerated rapidly into the distance.

Wayne suggested that they wait as the craft would return. It did. While the participants did discuss the sighting among themselves, no official reports were ever made. Due to the proximity of Sandia National Laboratories and Kirtland Air Force Base, their first assumption was that it was an experimental aircraft. Speed of departure seemed to rule that out as it exceeded anything possible at the time.

He indicated that no other cars passed them during the observation and landing. That seems highly unusual, as at that time Route 66, which was replaced by Interstate 40, was the major east-west highway in the United States. Wayne said that he talked to a USAF pilot who lived in his immediate neighborhood about the incident. The pilot was home during the day, which was unusual. Allegedly the pilot said he had been temporarily grounded for chasing a UFO. Wayne said that the pilot also saw beings through round portholes in the craft.

This case is provided to illustrate what can happen if you just ask. There are thousands of people, who like Wayne, will tell you about their experiences, provided they believe you are sincere in your interest. They have a legitimate fear of ridicule, and there are too many people who will undermine their confidence. Be a friend and you'll learn a lot by just listening.

Summary

When skeptics pontificate that there is no evidence to support the existence of UFOs, they choose to blindly ignore the vast amount of documentation that has accumulated over recent decades. As depicted in the cases covered in this chapter, there exist numerous high-quality eyewitnesses that have reported personal experiences, often supported by corroboration from technical means such as photography or radar. Especially noteworthy are the cases involving many witnesses, and on occasion at very close range.

Of great significance are the cases involving interruptions by UFOs of strategic military weapon systems, as well as their unwarranted endangerment of civilian airliners. The true extent of sightings is unknown as most go unreported. A major shortcoming is that there is no duly authorized central location or organization that is both willing and competent to register the data, let alone investigate worthy cases.

Other countries have been concerned about UFOs, but addressed the public quite differently. The following chapter discusses the approach of the United Kingdom.

THE UK CONNECTION

The evidence that there are objects which have been seen in our atmosphere,
and even on terra firma, that cannot be accounted for either as man-made
objects or as any physical force or effect known to our scientists
seems to me to be overwhelming.
—LORD HILL-NORTON, FORMER UK CHIEF
OF DEFENCE STAFF

It is important to note that other governments have handled the UFO topic
quite differently than has the United States. The government of the United
Kingdom offers an excellent example. A good deal more information has been
made public about the subject, largely due to the efforts of Nick Pope. He is a
former Ministry of Defence (MoD) official, who has written several books
about his experiences and frequently appears at both conferences and on television specials covering UFOs.

Meeting the MoD Chief Scientific Advisor

My formal encounter with the UFOs and the UK came in a situation that led to
considerable personal embarrassment. In 1992 while working at Los Alamos I
received notice that a very senior official from the British Ministry of Defence
wanted to meet with me. That was all of the information that was received.
That person was Sir Ronald Oxburgh, who was then the Chief Scientific Advisor to the Minister of Defence. Oxburgh held a Ph.D. from Princeton and his
résumé included being the head of the Department of Earth Sciences and
president of Queens College at Cambridge University. In the United States he
had also been a visiting professor at Stanford University, California Institute of
Technology, and Cornell. Obviously this was a very straight biography that
there was to go on.

In the early 1990s the vast majority of my efforts at Los Alamos were on
nonlethal weapons. That work had just begun to attract international public
attention and there were articles written about the topic that often included

my name. Some of the media attention included British publications and interviews with the BBC. Based on his background and current position, the reasonable assumption was that Oxburgh wanted to discuss nonlethal weapons.

Dr. Oxburgh arrived at the laboratory accompanied by a small entourage of British scientists. The meeting was held in a room in the headquarters building of LANL that is located "behind the fence." That means it was in a classified facility. In reality, there are very few secrets that we keep from the UK, including ones regarding nuclear weapons. The basic agenda included a series of technology presentations that lasted a few hours. After the briefings those scientists accompanying Oxburgh were taken on a tour of some of the more sensitive places around the laboratory complex. As Oxburgh had visited LANL several times before, he did not go with them. It was at this point that he and I met privately. Prepared was my typical Nonlethal Weapons 101 briefing, as it was assumed that was the topic he wanted to discuss. Nothing to the contrary had been indicated.

We were the only two people in the room and the briefing lasted about an hour. After that, informal discussions about the topic lasted another hour. Among things, he told me about problems they had encountered with smugglers in go-fast boats when Hong Kong was a British protectorate. It was a pleasant discussion that ended when his team returned.

About two weeks later I received a phone call from a person at CIA Headquarters. "You really blew that one, Alexander," I was told. At that point I didn't even connect the comment to Oxburgh. Further discussion revealed that the topic Oxburgh wanted to discuss with me was UFOs, not nonlethal weapons. I told the caller that since we had been the only people in the room, he could have changed the topic at any time. During the entire two-hour conversation I never had the impression that he was not interested in the nonlethal concepts that were currently being worked on, or that he wanted to discuss something else. Opportunity missed.

A short time after his visit to Los Alamos, Dr. Oxburgh left the MoD and was made a Knight of the British Empire whereupon he became Lord Oxburgh, also Baron Oxburgh of Liverpool, and afforded a Life Peer to the House of Lords. In 1993 he assumed his next position as Rector of Imperial College of Science, Technology, and Medicine in London.

In 1993 and 1994 I was again appointed to a NATO/AGARD study of nonlethal weapons. The study director was a senior UK MoD official who worked at Farnborough. That study took me to London, so a meeting with Sir Ronald was arranged at his office. After taking the tube to Kensington Station I walked the short distance to Imperial College. The security guard pretended that he didn't know who I was talking about when I mispronounced the name as Oxborough. When it was explained there was an appointment with the rector, he informed me of the correct pronunciation, which can't really be replicated in print as it must be executed with a quick exhalation.

This meeting went very differently from the one at LANL. It was very relaxed and we openly discussed UFOs. One of the issues that had caught his attention while at MoD was the problem of crop circles that had plagued British farmers for years. Oxburgh told me that while visiting various bases throughout the country, he had seen interesting photos of these agroglyphs that had been taken by aircrew members. These were not the pictures commercially available in UFO popular literature but ones they had taken when flying over the sites. He was interested in a range of topics, but clearly had no more information than I did. In fact, he was looking for answers, which indicates that UFOs were not an issue that made it through the filters to the senior leadership. Of course, one of the inferred questions regarding the crop circles was, *"Is this something you [the U.S.] are doing?"* While there was no reason to believe it was, this implication repeats the ubiquitous presumption—someone else is doing it!

UFOs, Nick Pope, and His Father

In response to Oxburgh's questions I suggested to him that he might want to get in touch with Nick Pope who had officially worked the topic in MoD and was then writing publicly about it. Lord Oxburgh's response was both memorable and important for perspective. He said, "Nicky? Why his father used to work for me as my deputy." Here was the former MoD Scientific Advisor, who personally knew Nick Pope, and yet he had no idea that he had been involved in UFO studies. That pronouncement actually speaks volumes about the relative importance that MoD placed on the topic. The most senior person in MoD in the UK involved in science and technology matters was not even aware that the study had been ongoing for years. What that says to readers is the UK MoD did not consider UFOs to be very important. If it was a hot topic, as some researchers have suggested, it would have made it through information filters to Oxburgh. After all, he held information about many of the most secret projects in the United Kingdom and the United States. Had UFOs been considered a serious science and technology issue by the MoD, Oxburgh would have known about it.

This information is quite consistent with what Nick Pope has reported in his writings and presentations. While Nick has been called the British Fox Mulder (from *The X-Files*), that is probably not an accurate description. Overlooked by UFO enthusiasts is that while he was assigned to the Secretariat (Air Staff) much of Pope's work had nothing to do with UFOs, and he had other duties he carried out in parallel. He held a grade of senior executive officer and his tenure spanned twenty-five years. His formal association with UFOs was for only a portion of that time.

Where most people have not connected the dots is from Nick Pope to his father, Dr. Geoffrey Pope, who was a very important figure in British aerospace development. Obviously, as the Deputy Chief Scientific Advisor in MoD, Geoffrey Pope was in a position to have access to almost all information about science and technology that was of significance.

Well before my Imperial College meeting with Oxburgh, Nick Pope and I had met previously near Whitehall and discussed our mutual interests. After my UFO discussions with Oxburgh, we met again and I told him about the exchange and what had unexpectedly been learned. I also suggested that he might expect a call from Oxburgh or take advantage of the opportunity and arrange his own meeting. It was then that Nick confirmed the prior professional relationship between his father and Lord Oxburgh.

Over the years Nick Pope and I have met on several occasions, sometimes at UFO-related conferences. We have been able to compare notes quite openly and our experiences in studying UFOs were very similar, despite the geographic and organizational differences. The congruent bottom line was that there was no major UFO program at any security level that we were aware of. We agree that there are a core set of cases that defy traditional explanations. Very significantly, we both agree that UFOs should be a national defense issue and represent a potential threat to commercial aviation.

And, of course, we have discussed his father. Geoffrey Pope was a career civil servant with an extraordinary background. Dr. Pope joined the Royal Aircraft Establishment (RAE) at Farnborough in 1958 as an aeronautical engineer and worked on some of the most vexing aviation problems. He held several posts in the MoD at Whitehall and returned to Farnborough as the director in 1984. This was a key position as Farnborough RAE was the leading British military aerospace research and development facility. Before becoming the MoD Deputy Chief Scientific Advisor, some of the other positions he held were group head of Aerodynamics, the Deputy Director (Weapons), Assistant Chief Scientific Advisor (Projects), and Deputy Controller and Advisor (Research and Technology). Dr. Pope was also the President of the Royal Aeronautical Society and worked extensively with the United States aerospace research community. In fact, he was a recipient of the U.S. Secretary of Defense Medal for Outstanding Public Service. The point is, Nick

Pope's father was very much an insider when it came to aerospace research and development.

In our discussions, Nick Pope mentioned that he had talked to his father about UFOs. Among the topics was whether or not a crashed craft existed. The response as told to me was quite impressive, and probably accurate. Geoffrey Pope said to his son, *"If we had a UFO it would have been in my hangar."* He meant by that it would have been located in the extensive research facilities at Farnborough. Clearly it wasn't.

UK MoD Decide to Release Their UFO Files

In 2008 the UK MoD made a conscious decision to make all of the previously classified UFO files available to the public. Prior to that time they employed the U.S. model of data release by responding only to FOIA requests about specific cases. They were well aware of the French experience in releasing documents over the Internet. In 2007 the French space agency, Centre National d'Etudes Spatiales (CNES), made an announcement that all files could be accessed by the public as of March 23 of that year. Over the years CNES had accumulated more than 100,000 pages of material based on reports of more than 1,600 sightings. The unanticipated response was overwhelming; so much so that the influx of people attempting to get at the data crashed the entire site within three hours.

The move to release files was more pragmatic than altruistic. There were four reasons provided for making the release in the manner they chose:

— Too many FOIA requests regarding UFOs
— The CNES files had been released by France
— A demonstration of open government and transparency
— Defuse accusations of a UFO cover-up

The United Kingdom, just like the United States, was inundated with FOIA requests asking for information about UFOs. In fact, it was the most requested topic. Like their U.S. counterparts, when enacted the founders of FOIA never saw that one coming. The administrative burden was more than confidentiality was worth. Of course, they needed to redact some files to protect personal privacy.

The MoD wisely decided to do a timed release rather than dumping all of their files at once. The first batch, made available on the Web site of the UK National Archives in May 2008, covered the reports from the years 1978 through 1987. A BBC article quoted a briefing prepared by MoD for the House of Lords for a debate about UFOs that sounds very much like what we would expect from the military establishment in the United States. The briefing said that "there is nothing to indicate that UFOlogy is anything but claptrap" and that the idea of an "inter-governmental conspiracy of silence" was "the most astonishing and the most flattering claim of all." The briefing goes on to say: "Let me assure this House that Her Majesty's government has never been approached by people from outer space."

Although it had been previously disclosed, among the reports released in February 2010 was a 1952 Churchill memorandum in which the World War II Prime Minister, Sir Winston Churchill, inquired about the topic to Lord Cherwell, then Secretary of State for Air. Churchill's note stated, "What does all this stuff about flying saucers amount to? What can it mean? What is the truth?" The response to him indicated that everything could be explained with prosaic answers including meteorology, astronomy, misidentified conventional aircraft, delusions, and deliberate hoaxes. The response also indicated that the Americans had reached similar conclusions. There was no mention that there were a few cases that defied explanation.

Then in August 2010 a few more UFO-related files were released by the UK MoD. Unfortunately, mainstream news media in Europe and around the world filed ridiculous stories with titillating, but knowingly misleading, headlines. As an example, in the UK the *Telegraph*'s banner read "UFO files: Winston Churchill 'feared panic' over Second World War RAF incident." Without qualification, the *Toronto Star* claimed, "Churchill ordered cover-up of encounter between spy plane and 'unknown object.'" In America *CBS* turned the issue into a question: "Winston Churchill Ordered UFO Cover-up?" They were not alone, as *ABC* and *Fox News* posted similar stories. General Eisenhower was mentioned in several articles, and *The Asian Age* headline ran, "Churchill, Ike covered up UFO encounter." Such stories were indicative of extremely unprofessional reporting of facts, as well as a blatant attempt to manipulate the topic merely to attract readers.

In reality, the sole basis for those news reports was a letter of inquiry sent to MoD in 1999 by a man who reportedly heard a story from his mother, who said her father told her that he had overheard a telephone conversation to which he was not a party. Further, the incident allegedly happened more than fifty years earlier and was totally lacking in supporting data. Such reporting of unsubstantiated fourth-hand information as fact is demonstrative of the disdain in which the media in general holds the topic. It is also indicative of the media's low regard for the intellect of their readership.

Fortunately the MoD later took a broader look at the issue. As with many other studies, about 5 percent of the cases cannot be easily explained. A 5 to 7 percent remains amazingly consistent across most of the studies. The vast majority have conventional determinations, often misidentification of known objects. Nick Pope has indicated that at least a few of the reports were considered to be problematic. Those cases came supported by military radar reports, video that was examined and found to be unadulterated, and highly skilled, competent observers. Those witnesses often came from law enforcement or the military itself. Of interest were sightings in which the UFOs performed aerial maneuvers that cannot be matched by current state-of-the-art military aircraft.

There Were a Few Convincing Sightings

As an example, in March of 1993 a sighting occurred with more than sixty witnesses, including military personnel and a meteorologist. The latter observer is rare but important as skeptics frequently suggest that weather anomalies cause both unusual sightings and radar malfunctions. The witnesses described a large triangular object that was larger than a Hercules C-130 aircraft. At first the UFO moved very slowly, but suddenly it shot off at high speed. It is this combination of flight characteristics that make these craft so unusual. A report was forwarded by Pope's boss to the Assistant Chief of the Air Staff. It concluded: *"It would appear that some evidence that on this occasion there was some unknown object, or objects, of unknown origin operating over the U.K."*

There is a long and rich history of UFO sightings over the United Kingdom. Some were released long ago and at least one became one of the most puzzling cases in both Blue Book files and the Condon Report. This involved a 1956 sighting near Lakenheath, UK, which was supported by both radar and visual sightings by two Venom interceptor pilots that were vectored to the area. The case was so spectacular, yet controversial, that Gordon Thayer wrote an article about it in the *Journal of Astronautics and Aeronautics* in September 1971. The event began on the evening of August 13 with returns on the Bentwaters GCA radar. The speed is not certain but 4,000 miles per hour was the slowest estimate. Some estimates were in excess of 10,000 miles per hour. Not bad for 1956. These returns were so unusual that checks were made to determine if the radar was malfunctioning. It appeared to be operating correctly. About five minutes later a group of twelve to fifteen unidentified targets was picked up, but moving at about 80 to 125 miles per hour. The UFOs then merged into a single source that was described as several times larger than a B-36. The radar echo remained stationary for ten to fifteen minutes, followed by movement of a few miles, and then resumed its stationary pose.

A short time later the UFO was again spotted moving in excess of 4,000 miles per hour. Bentwaters control then contacted Lakenheath and asked if they had spotted anything moving at 4,000 miles per hour. They noted that a C-47 pilot had reported something streaking by but could not make out what it was. Shortly Lakenheath radar spotted the UFO, which was again almost immobile. However, it then began moving at 400 to 600 miles per hour. Of note was that the UFO went from zero to that speed instantly; it did not follow anticipated increasing acceleration. After more than half an hour the RAF scrambled a de Havilland Venom night fighter and they were vectored toward the UFO. The pilot flew over Lakenheath and indicated he could see a bright white light. The pilot also advised that he had radar contact and was "locking on." The Venom indicated he had his guns locked on the target. He later told investigators this was "the clearest target he had ever seen."

Almost instantly the target disappeared. The Lakenheath radar operator informed the pilot that the UFO was now behind him. The pilot attempted to evade the bogey but couldn't shake it. Reporting he was low on fuel, and obviously shaken, the Venom pilot headed back to his base. Lakenheath radar reported that the UFO followed him for a short time, then returned to a stationary position. Investigators at the time stated they believed this was a real target. Worthy of consideration is the timing of the disappearance of the UFO. It was when the Venom pilot indicated that he had a lock on the target. Other than that circumstance, the UFO seemed to be willing to be observed. It was clearly in control of the situation.

Comparison of these cases that occurred almost forty years apart yields similarities that cannot yet be accounted for. The UFOs both moved very slowly and yet were able to accelerate to speeds that are unmatched by any known conventional aircraft. There were both multiple radar observations and credible eyewitnesses. Equipment checks confirmed the radars were operating properly. Interestingly, the UFO nemesis, the late Phil Klass, just wrote off the 1956 case as radar malfunction—even though the facts suggest otherwise.

An American Pilot Is Ordered to Fire on a UFO

There was another case that was released in the second batch of MoD reports that deserves special attention. While the report came out from London, it involved an American F-86D Sabre jet pilot, then Lieutenant Milton Torres. Like Nick Pope, I have had a chance to meet Torres, who is now a retired mechanical engineering professor with a doctorate from Florida International University. Both Pope and I agree that Torres is extremely credible, and I will return to other aspects regarding the personal impact of this case on him later in the book.

The important confrontational issues involve engagement against a UFO on May 20, 1957. Lieutenant Torres was assigned to the 514th Fighter Interceptor Squadron of the 406th Air Expeditionary Wing at RAF Manston in Kent. That night he and another pilot were on strip alert and with five minutes notice were ordered to take off immediately. Uniquely, they were instructed that this was not a drill, and it was a hot mission. Sent to protect the United Kingdom and the rest of Europe from a potential Soviet attack, this was the first time he had ever received such instructions.

Torres and his wingman were told that the air defense radars had been observing a UFO that was exhibiting very unusual flight patterns. At times it would remain motionless for long periods. They were told to accelerate to maximum sustainable speed, which with afterburners was about Mach .92. Concerned about his orders, and understanding the gravity of the situation, Lieutenant Torres requested an authentication code and received the appropriate response. This was not a drill. The Sabre jet was equipped with 24 2.75-inch Mighty Mouse Mk4 rockets, which while not very accurate, any one rocket did have the explosive power to bring down an aircraft. Approaching the UFO, Torres reported the size of the target, its return at fifteen miles, and suggested the object "had the proportions of a flying aircraft carrier." He was fully locked on to the target and preparing to salvo all of his rockets, which would ensure a hit. Then suddenly the UFO took off at amazing speed. Torres estimated that it went from stationary to Mach 10 almost instantaneously. He noted the UFO executed a right-angle turn at speeds that would induce unsustainable g-force on any human pilot. Those capabilities convinced Torres that the craft was indeed of extraterrestrial origin.

The Royal Air Force ground radar also tracked the departure. The station indicated the UFO had exited the scope, a range of 250 miles, in two seconds. These multisensor observations suggested the UFO was exceeding the known laws of classical Newtonian physics.

Upon his return to base, Torres was ordered to not mention the incident ever again. He was a young pilot, and some interrogator threatened him by stating that if he did talk, he would lose his flight status—a fate worse than death to a fighter pilot. He did not say another word about the case. During Vietnam he flew 276 combat missions and was awarded the Distinguished Flying Cross. He retired from the U.S. Air Force in 1971 as a major. According to Torres, the person who interviewed him waved an NSA badge, but he has no idea whether or not that was valid. Therefore, he was caught by total surprise when Nick Pope called him in October of 2008 and gave him twenty-four-hours warning that the entire incident was about to become public knowledge. As far as is known, the case has not appeared in any American files.

Worth commenting on is a case involving a commercial airliner and a near-miss incident. This event supports our comments about air safety concerns. On April 21, 1991, an Alitalia MD-80, with fifty-seven passengers on board reported a close encounter with a UFO. The description was of a brown cigar-shaped object that flew so near the airplane that the pilot shouted out instinctively. Pope indicated that his investigation eliminated all of the usual suspects such as weather balloons or military aircraft. The incident took place at an altitude of 22,000 feet over Kent. The object remains unidentified but the event does point to an air safety issue.

There are many other sightings over the United Kingdom but few that have the multisensory confirmations that support credible eyewitnesses. The case that has gained a reputation as the UK's Roswell, Rendlesham Forest, has been briefly addressed earlier, and will be handled in more detail later.

The UK Has Its Own Investigation: The Condign Report

Given the number of incidents that had been documented in the United Kingdom, and the timing of them, the Condign Report, sponsored by the Defence Intelligence Staff between 1997 and 2000, came to some rather re-markable conclusions. While called *The Condign Report,* the researcher could probably have saved time and money by repackaging the U.S. Condon Report done thirty years earlier. The study also seems to have taken lessons from the U.S. Air Force's inept attempts at explaining Roswell. The bottom line in all such studies seems to be to disregard the facts and allow any expla-nation except the one that is glaring them in the face and is already believed by the public.

Like Condon, Condign indicated that unidentified aerial phenomena, or UAP, were not a threat. As the U.S Air Force found, removing a potential threat context defers responsibility. They also followed Condon in noting that maybe these were somebody else's aircraft, and they just didn't know about them yet. There is a nod to U.S. advanced aircraft that are still secret but are allowed to use UK bases. This is partly feasible as mutual defense pacts do ex-ist between the countries.

In the U.S. Air Force account explaining alien sightings at Roswell, they stated they were anthropomorphic dummies used in high-altitude parachute tests. The problem was that those tests did not occur until six years after the 1947 event. Condign does a similar temporal transformation. While they may explain a few cases in recent years, they certainly do not answer the observations from the 1950s that we just covered.

While the events at Rendlesham Forest must have been available to the study group, certain known evidence was obviously rejected or ignored. Specifically Condign indicates that there were not any trace-radiation cases, a fact directly refuted in publicly released reports from the Rendlesham case. We also know that casts of imprints were made at that site. In the Condign Report great effort was taken to infer that natural phenomena, such as meteors and atmospheric conditions, were responsible, even after they were categorically discounted in the initial investigation. Unexplained observations were relegated to phenomena that were not yet understood. Obviously disregarding evidence from cases such as Torres and Lakenheath, the report states there was no evidence of control. The response really points more toward how large institutions deal with unpleasant topics that they wish would go away.

The Crop Circle Puzzle—More Complex Than Meets the Eye

Crop circles have a historic association with UFOs though a causal relationship is far from established. Nonetheless, as noted by Lord Oxburgh, there has been considerable interest in speculation about their origins. There is no doubt that some of these agriglyphs are man-made, especially after so much fuss began to be raised. There is also no doubt that a couple of drinking buddies with too much time on their hands did not go out and make some of the very sophisticated fractal designs. Also, given that crop circles are found in many countries, their frequent flyer mileage should give them away. Of course there are always the imitators who quickly acquired nocturnal circle-making skills and entered the field.

In some cases the size, complexity, and speed with which the agriglyphs appear to have been made are problematic for simple conventional explanations. There are a few instances in which lights from unknown sources have been both observed and photographed over the fields. Some research indicates that demonstrable changes in growth patterns have been reported. These changes would not be accounted for by pressure from a human's foot. Overall, the connection between crop circles and UFOs is tenuous at best. They do, however, share similar qualities in that neither phenomenon has any easy answers that fit all of the facts.

One hypothesis that has been put forth by technologically savvy scientists is that the designs are created with a beam weapon operating from a stealthy airframe. While they believe the origin is U.S., this craft obviously would not be an F-117 or B-2, the acknowledged American stealth aircraft. Had this hypothesis not come from scientists I know and trust, the concept would not have even been addressed in this book. Still, I find the hypothesis extremely improbable for several reasons. Worth noting is that my research and development background did include considerable work on directed energy and beam weapons and the state of the art as of a few years ago was pretty well known. What the scientists have described is well beyond anything I am aware of at this time. To support their allegations, the secret beam weapon would have to have been operational from space or a stealthy aerial platform for years to decades. That is not the case.

There are a host of conceptual problems with the application, even if this weapon did exist. First, it requires a stealthy platform, one not observable in the visual spectrum or picked up by conventional radar systems. The data provided suggested that it would have to operate at about 20,000 feet, which would raise air safety issues. Next, one must answer why we would invade one of our closest allies in what could be considered an act of war?

The strategic implications of getting caught would be enormous. After all, the United Kingdom is the one country that has always been reliable in the U.S. GWOT operations. For the United States military to be conducting exercises that anger British farmers is unthinkable. In Pakistan and Afghanistan collateral casualties from air-to-ground missiles have been a huge political albatross. Yet there are people who suggest that we have a directed energy weapon with pinpoint accuracy but don't use it. Obviously, the question of why such an advanced system would be used to draw interesting diagrams in crops, rather than being employed in our wars, begs a logical explanation.

The UK MoD Decides to Shut Down the UFO Office

For many years the UK MoD had an office that officially investigated UFOs. Many readers will be familiar with the books and television appearances of Nick Pope, whom we've mentioned held that position for several years. However, remarks that were redacted on released reports, including derisive commentary from other officers who were stuck with UFO investigations, suggest that Nick was indeed a rare commodity in the MoD. Obviously not all of those assigned to those duties were as impressed with the information as he was. A 2007 memo to ministers and defense chiefs noted that, "The files are considerably less exciting than the 'Industry' surrounding the UFO phenomena would like to believe." It also stated the MoD position quite clearly: "contrary to what many members of the public may believe, MoD has no interest in the subject of extraterrestrial life forms visiting the UK, only in the integrity and security of UK airspace."

On November 11, 2009, the Ministry of Defence announced that the office was being shut down, even though 634 sightings were reported just that year. That is well above the norm for the past decades, and in an economic environment of restricted resources was seen as detracting from serious MoD activities. There was, however, a more significant action that accompanied the MoD directive that "these facilities be withdrawn as soon as possible." That was the order to destroy all new reports of strange sightings within thirty days. Just as noted with the American CIA blocking new UFO reports, the MoD, after releasing their holdings, wants to ensure they do not have to keep searching the files in order to respond to public inquiries into the subject. This action should also serve notice as to just how important the UK government believes the topic is. As in the United States, it seems that British UFO enthusiasts have succeeded in killing their golden goose as well.

Summary

While the UK Ministry of Defence made a conscious decision to proactively release their UFO data, it is clear that the scientific consensus mirrors that of the United States. They remain relatively skeptical, despite the few dramatic cases that were found in their files.

More important may be what the senior UK MoD officials did not know. The United States and the United Kingdom have a special relationship, especially in scientific and security matters. As evidenced by the people mentioned in this chapter, and despite their personal interest in the topic, there is no indication that they see the situation any differently than do our senior government officials.

It was also interesting to note what happened when the official files were released to the public. There was high interest, as indicated by the number of people accessing the Web site. Certainly the case of Milton Torres directly addressed the existence of craft far beyond current human capabilities. However, disclosure of that information did not bring about any civil discord, as has been hypothesized by UFO buffs.

As noted in the case of Lieutenant Torres, sometimes secrecy was invoked regarding UFOs. The following chapter examines just how inconsistent that process was.

SWORN TO SECRECY

Secrecy, once accepted, becomes an addiction.
—EDWARD TELLER

The History

The issue of secrecy surrounding UFOs is probably one of the most abused and misunderstood aspects of the UFO narrative. There is little doubt that when the subject first came up, the U.S. military did highly classify almost everything related to UFOs. The senior leadership of the country did not know what was going on and were loathe to admit it.

My experience with UFOs and secrecy is probably quite different from civilian expectations. Of course conspiracy theorists will simply discount these comments as part of the cover-up. That is not true, and as stated in the prologue, no misleading statements will be published. What I have learned over the years of studying this topic is somewhat different from what I expected. In fact, as discussed in Chapter 1 on Advanced Theoretical Physics, initially members of the group, including me, believed that some black organization did exist and was handling UFO data.

Given my position when on active duty, I was able to see some of the original classified documents as well as the stories being printed in the open media. In comparing the two, what was determined was that in most cases about 98 percent of the information was already in the public domain. That included all of the critical information about the UFO events themselves. The remaining classified portion had to do with what is known as *sources* and *methods*. Sources and methods are the techniques or tradecraft secrets that are used to obtain information. The reason they are protected is that these same techniques may be used on many occasions to collect information about a wide range of topics.

As an example, for a long time the public was not aware of the government's ability to tap into cell phones and track the physical location of the user even if the device was turned off. In both intelligence circles, as well as in law enforcement, the general public's ignorance of that technical capability was useful. The Intelligence Community (IC) collected information surreptitiously and the police were able to find bad guys with great efficiency. As the word finally got out, both terrorists and criminals began to use alternative measures. Stolen cell phones and cheap models are frequently used for a brief period and then thrown away. Sometimes they are placed on a moving conveyance, just in case they are being tracked at that moment. Terrorists and organized crime figures are highly adaptive. As soon as they learn of the technical capabilities they change their modus operandi. It is for those reasons that sources and methods are so highly protected—and appropriately so.

One of the famous cases in which some official reports remained classified for a time was the incident over Iran on September 19, 1976. This is one of the strongest cases that demonstrate why there should be defense concerns about UFOs. The data included multiple ground and aircraft radar observations plus many highly credible eyewitnesses. What is so important was the capability of the UFO to negate Iranian Air Force weapons and block communications—in aircraft that had been provided to them by the United States.

At about 12:30 A.M. responding to calls from civilians in Teheran, operators in the control tower at Mehrabad Airport saw what appeared to be a bright light that could not be explained. The tower operator called for a response from Shahroki Air Force Base, and two F-4s were scrambled at about 1:30 A.M. The light was so bright that the first pilot, Captain Mohammad Reza Azizkhani, spotted it from a distance of more than seventy miles. As the first pilot approached within twenty-five nautical miles of the object he lost all instrumentation and communications capability. The pilot turned away from the UFO, and a short time later regained the electronics that had been lost. Ten minutes after the first fighter took off, it was followed by a second F-4, piloted by Lieutenant Parviz Jafari. The lights of the object were intense, making the size difficult to determine. They were alternating blue, green, red, and orange, and were arranged in a square pattern that flashed in sequence, but the flashing was so rapid that they all could be seen at once.

The second F-4 acquired a radar lock on the object at 27 nautical miles range. The radar signature of the UFO was reported as resembling that of a Boeing 707 aircraft. As Jafari's aircraft was closing on the object at 150 nautical miles per hour and at a range of twenty-five nautical miles, the object began to speed up, maintaining a steady distance of twenty-five nautical miles from the F-4.

While the object and Lieutenant Jafari continued on a southerly path, a smaller second object detached itself from the large UFO and proceeded toward the F-4 at high speed. Lieutenant Jafari, thinking he was under attack, tried to launch an AIM-9 sidewinder missile, but he suddenly lost all instrumentation, including weapons control, and all communication. What has been unclear in all of the reports is whether it was after Lieutenant Jafari activated the weapon system or when he thought about shooting that the weapons panel was disabled. Lieutenant Jafari then instituted a turn and a negative g dive as evasive action. The object fell in behind him at about three to four nautical miles' distance for a short time, then turned and rejoined the primary object. The main object then flew away at several times the speed of sound according to a voice tape from one of the pilots.

Once again, as soon as the F-4 had turned away, instrumentation and communications were regained. The F-4 crew then saw another brightly lit object detach itself from the other side of the primary object and drop straight down at high speed. The F-4 crew expected it to impact the ground and explode, but it appeared to come to rest gently on the ground. A search of the dry lake area the next day failed to reveal anything.

In addition to the interaction with the Iranian Air Force interceptors, a civilian airliner that was transiting the area at the time of the incident also reported loss of communications when approaching Mehrabad Airport. Years later, operators who had been in the tower that night stated a UFO had come over them and knocked out power there as well. What was most significant to me was not the UFO's ability to shut down electronics; we know how to do that. What was more interesting was its ability to restore power when there was no longer a perceived threat.

Lieutenant Jafari stayed in the Iranian Air Force and became a lieutenant general. He has traveled to the United States and discussed the incident publicly. Others involved have also made statements. The case even became a television special and was incorporated in several UFO-related programs. It stands out because of the documentation available. Not surprisingly, Phil Klass, again ignoring the facts, attempted to write it off as a misidentified celestial object and incompetent pilots, and poorly maintained aircraft equipment.

The amount of cooperation between the Iranian and the U.S. Government in the investigation is not certain. It is likely that part of the reason for the initial classification was so that the then friendly government of Iran would not be aware of the ability to get information about their internal activities.

The Devastating Psychological Impact Secrecy Can Have

In the last chapter the case of Milton Torres and his attempted interception of a large UFO was presented. As noted, in that case the UFO simply departed, accelerating at an astonishing rate. But that is not the end of the story regarding the security aspects of the case. When Torres returned to his base he was instructed to call in on the landline as soon as he was on the ground. He was told not to say anything to anyone about what had transpired that night. The next day he was debriefed by someone whose identity was, and remains, unknown. At the end of the debriefing Torres was admonished to remain silent. If he failed to follow those instructions, the debriefer claimed Torres would be taken off flight status, which would have been devastating to a young pilot.

For half a century Torres remained absolutely silent about the matter. Never again was he contacted, until Nick Pope called him to tell him that the case was about to be made public. What is dreadfully underestimated in these secrecy oaths is the potential for emotional trauma. I have heard Dr. Torres describe his experience publicly on multiple occasions. Every time he has broken down in tears due to that emotional impact. In particular, he laments that he never told his father of the event, which he considered absolute proof of extraterrestrial life. Unfortunately his father died before Torres knew that he was no longer bound to secrecy. The intense psychological burden of the invocation of secrecy was tremendous, and continues to this day. Worse, the requirement was probably unwarranted.

One of the great shortfalls in the intelligence system is the lack of follow-up. When one is sworn to secrecy about projects or events, it is rare when word is sent regarding declassification. There are a number of projects that I don't discuss, even after seeing articles written on the topic. The problem is that I have no way of knowing if the projects were formally declassified, or if some writer was just speculating on the topic. In addition, after one leaves the system, for that reason there is almost no procedure to get such clarification regarding continuation of classification. Few of these secrets have the impact of the one described by Torres, but for many operators there is an emotional impact and a price to pay—often needlessly.

One of the most interesting books related to UFOs and secrecy is *UFOs and Nukes* by Robert Hastings. In his book he details many, many cases of UFOs interacting with nuclear weapons bases, laboratories, and reactors. What was most impressive about the book was the detailed information that former military personnel provided him. Of significance is the number of people who are clearly identified by name. When not specifically named, enough information was provided that any internal investigator would be able to make the connection if so inclined. Hastings specifically addressed the issue of formal secrecy and noted that none of his sources had ever been questioned, reprimanded, or threatened over their disclosure of encounters with UFOs.

Hastings came to the same conclusion that I did some time ago. That is that there are some people, such as Torres, who were told to remain silent and some of them were even threatened in the process. However, it firmly appears that these were the efforts of low-to midlevel agents who were doing what they thought was needed at the time. This is very different from an official uniform policy that emanates from a duly appointed authority. The agents most likely had heard all of the rumors, much like the civilian population. Encountering a UFO investigation would be a very low-probability event, thus rarely experienced. Therefore, as is too frequently the case in the Intelligence Community, it is easier to invoke secrecy than to clean up a breach after it has occurred.

In most of the cases that I have investigated that involved the possibility of formal inquiry, I have asked the participants if they ever were debriefed, and/or told they could not talk openly about the event. None of them had, though sometimes they had heard of second parties that had been told they should remain silent. Among the cases I refer to are some of the most significant UFO interactions between military systems and UFOs. That includes the participants in the Malmstrom missile shutdown and Bentwaters. It is important to note that although both of those cases involved strategic nuclear weapons, there were no admonitions reported when witnesses did reveal their information.

Keeping Secrets

There were those who did keep secrets. Colonel Bill Coleman was one of the important players. In fact, he was also responsible for some of the rumors about UFOs and happy to hear them. Bill is a truly unique individual and his career spanned many decades. Entering the Army Air Corps as a teenager during World War II, like many men of "the Greatest Generation," he abruptly transitioned from childhood to adulthood. While assigned in the Pacific Theater he flew P-38 Lightnings and participated in devastating death-dealing strikes against Japanese forces trapped in the Philippines. Deserting the farm, he stayed on active duty after the armistice was signed, and later participated in operations in Southeast Asia during the Vietnam conflict. He was also involved in observing atomic tests, and ended up paying quite a price. A few years after the explosions, Bill was diagnosed with leukemia, a disease that felled many of his comrades who had been present at those tests. However, thanks to some unique medical interventions that he can't even explain, he has been in remission since 1958.

In the late 1960s, Bill was nominated to become the Chief of Public Information for the U.S. Air Force. Due to the sensitive nature of the position he was interviewed personally by Secretary of the Air Force Gene Zuckert. Interestingly, the topic of UFOs was brought up. Since he would be working directly under the Secretary and be an advisor to him, Bill informed him that he was not neutral on the subject. There are not many Air Force pilots who admit that they have been involved in a chase of a UFO at such close range, but in fact, in 1955 he had an encounter with a UFO that remains unexplained.

Coleman's Amazing Encounter

His story went that at the time he had command of two squadrons of mixed varieties of aircraft at Greenville Air Force Base, in Mississippi. On the day of the event Bill had flown a B-25 Mitchell bomber to Miami and picked up one that had just been overhauled. In order to maximize his time he asked two technical representatives, one from Lockheed and the other from Allison, to accompany him. The purpose was to be able to discuss business during the rather long flight. After departing Miami, Bill turned the controls over to the young West Point copilot, pushed back his seat, and began the technical discussion. About ten minutes east of Marianna, Florida, the copilot called Bill and asked him about a strange object that was above them at an estimated 25,000 feet. At first Bill thought it to be a craze on the windscreen. However, as he took over the controls and maneuvered to the left, it became clear that it was a physical craft, not an illusion.

In a letter to me Bill picks up the story:

> If it was an aircraft it would normally be moving much faster as we were roughly on the same course. Suddenly it started descending. I asked the two reps to come up and see what we were seeing. I asked the flight engineer to move up into the bombardier's position, put on his earphones, and give us another viewing angle.
>
> We were now overtaking whatever it was. We all noted it did not look like an airplane since it had no tail or wings or other projections. We were now at the same altitude with it continuing to descend. I followed suit and advanced power to "maximum continuous military power." We were soon indicating 300 knots. We kept descending. Soon we were on the treetops and closing fast. I was getting confirming information from the flight engineer in the bombardier's position. I asked everyone to continue to grasp every detail they could.
>
> When we were within a quarter of a mile I told the crew, "Take a look at the shadow on the ground." The shadow was a perfect circle! I noted that I was going to overtake the unidentified. So, I

informed everybody that I would be making an abrupt hard turn right and then hard left turn and come up alongside and up-sun (for visibility advantage) of whatever we were chasing. We were now within one-eighth of a mile when I made the violent double turn . . . which should have placed us alongside the object.

When I made my final rollout there was nothing there! I immediately started a fast ascent to 2,000 feet. I asked everybody to keep their eyeballs out looking for our new "friend." Nothing! I then spotted a freshly plowed field of Alabama yellow soil to the left . . . approximately 600 acres (I'm a former farm boy). I yelled for everyone to look. The object was extremely close to the ground moving due north . . . leaving dual vortexes of yellow dust! I dove the Mitchell down to the deck to get behind (yes, behind) the trees so whatever it was couldn't see me.

When I came to what I figured to be the end of the field, I swung hard left in a 90-degree turn and expected to intercept our object at the end of the field. Alas, it had disappeared. However, I pointed out to everyone the existence of the two yellow vortex trails across the field . . . south to north. We continued to look but saw nothing.

The crew did see a bogey back at about 25,000 feet. Then the next day Bill had each of the people present write up the details of what they had experienced. He was explicit in that they were not to compare notes before submitting their reports. The report was given to the base intelligence officer at Greenville Air Force Base who indicated they would be forwarded.

Interestingly, when Bill had access to the Blue Book files, he never found his report. Unfortunately that is indicative of how many reports were handled. If the officer receiving the report did not deem it of sufficient importance, it just got dumped. There are many extra duties that are assigned, especially to lower-ranking officers. In general these are considered a pain in the butt. The unkind term for an officer stuck with these functions was SLDO. That stands for shitty little details officer and is not considered a good thing. You can imagine just how much emphasis gets placed on those jobs.

Confiscating Film

Among the incidents he participated in was what some people know as the missing Gemini photographs. These were pictures taken by Astronaut Gordon Cooper during his 1965 flight. According to Cooper, he inadvertently snapped some pictures of Area 51 while testing a camera. Colonel Coleman acknowledges that he is the one who confiscated the film, but only after quite a confrontation with NASA executives and Cooper himself. According to Bill, Cooper intentionally took the photographs and then tried to claim they belonged to him personally as he bought the film and took it along. At the time the base was extremely sensitive and very few of the senior executives were cleared to have any knowledge about it.

It took two trips by Bill and some rather tense discussions before the film was recovered. At that time there was no love lost between NASA and the U.S. Air Force. Despite grandstanding by Cooper, a terse call from the Pentagon ended the matter, but left most of the observers in the room in the dark about what was really on the film. The rumors of UFOs were easily acceptable to all concerned. For the record, Jim Oberg has addressed this matter as well. He claims that the orbit taken by Cooper on that flight would not have put him in position to photograph the secret base. That said, Bill did take the film and thus contributed to the lore.

Coleman was in one of the best positions to observe the Blue Book reports. He does indicate that out of more than 12,000 reports there are 105 that he calls worrisome. By that he means that those cases had two characteristics—high strangeness and high credibility. Those numbers go back to the 1960s. He claims he has the files and plans to write about them. Since he is now a widower in his mideighties he was urged to do it sooner rather than later.

A question that has arisen in some sectors is whether there was ever a connection between Air Force Colonel Bill Coleman and Army Lieutenant Colonel Phil Corso. Given their various assignments, there would have been some overlap in assignments at the Pentagon. Coleman says he never met Corso while they were on active duty or later, even though they lived less than one hundred miles apart in civilian life.

Secrets and Power

Despite various publications that have attempted to explain how the classification system is supposed to work within the U.S. Government, it still remains mystical to most civilians. A greater problem is that the system is grossly abused internally and that has nothing to do with UFOs. It does, however, lead the conspiracy theorists to wrong conclusions—many of which may seem almost logical given the paucity of facts available.

In Federal bureaucracies, as in many other organizations, information is power. Classified information creates its own mystique, and takes on an air of importance that may or may not be warranted. Unfortunately, there are some bureaucrats who believe that unless a document is classified, it is not important; of course, the higher the level of classification the better. The other side of the coin is that there is so much overclassification that real secrets are hard to sort out from the mundane. When on active duty I was aware of senior executives who refused to read the Black Book, which was a compendium of highly classified reports on current events. The problem, they noted, was that they could not distinguish between the Black Book reports and what they read in *Newsweek* or *The Washington Post*. One was required to initial a page when they read the book, thus affixing responsibility. By not reading it they could not be accused of accidentally leaking some classified information. In fact, it was rare when there was something that was truly important, and most of the classification was based on the analysis of the information rather than the raw data.

Secrecy is both expensive and can have negative consequences. As former Senator Daniel Patrick Moynihan wrote in his 1998 book appropriately entitled *Secrecy*, "The actors involved seem hardly to know the set of rules they play. Most important, they seem never to know the damage they can do." Germane to the UFO topic he wrote, "Conspiracy theories have been a part of American culture for two centuries. But they seem to have grown in dimension and public acceptance in recent decades." In my view he is correct on both counts and if the participants don't fully understand the game and consequences, how is the general public to be expected to comprehend it? The UFO conspiracies that abound are symptomatic of far larger problems.

A Rumsfeld Example

There are items of information that deserve to be classified, especially if they place troops in harm's way or degrade our national security. Unfortunately, the military and Intelligence Community tend to take the concept beyond common sense. As an example, a few years ago I was attending a meeting of staff members from the Office of the Joint Chiefs of Staff. They were conducting a study to respond to one of Secretary Rumsfeld's "snowflakes." These snowflakes were memos that engulfed subordinates, asking a myriad of questions daily. This question referred to the Global War on Terror and simply said, "How do we know if we are winning?" While attending the meeting I read portions of the GWOT promulgating documents that were classified Secret. I asked what was considered secret about the contents and noted, "If this is Secret, *The New York Times* must be Top Secret." The response was, "It is easier to classify everything than it is to figure out what really should be classified." While some will say the Bush administration was paranoid, not much has changed under Obama.

A major problem is that once a document is classified, the process to get it declassified is not simple. To classify a document, a person, known as a derivative classifier, just needs to wave a proverbial wand and the document becomes classified. Of course it is supposed to meet certain criteria, but those are not closely watched. To declassify that same document requires a review. That derivative classifier cannot just wave the wand again.

The issues are cultural, and simply adding another layer of bureaucracy, like the Director of National Intelligence, or creating the Department of Homeland Security, won't fix it. While institutional bureaucracy is not the topic of this book, it really does play an important role in understanding what is known and not known about UFOs. Secrecy is where the bogeyman lives. What is not known invites speculation. In this age of ubiquitous information, rumors spin out of control in an instant.

Secrets Not Stored

There is also misunderstanding about what happens to classified material. Documents adjudicated with a lower than Top Secret rating are generally stored in bulk. That means there are rarely any tracking numbers. In addition, there are many classified working papers. These are unfinished documents or even notes that may be accumulated in preparation of formal reports. When it comes to destruction of classified documents, there are usually no records kept of those classified Secret or lower. Before electronic storage came along, Washington, D.C., would have sunk in the swamp it is built on if all of that paper had accumulated. Therefore, there were mandatory destruction dates designed to get rid of bulk. When clearing out files to meet requirements, officers were credited by the linear feet of documents destroyed. That means we measured the space in each safe that was emptied. No record was kept of the subject matter in those files.

For UFO buffs, that means a lot of material that they are looking for simply doesn't exist. Previously mentioned was Howell McConnell, who worked at NSA and is known to some UFO researchers. McConnell did have a small number of highly sensitive files regarding UFOs in his possession. In some cases they had passed their normal date for destruction. Had he not kept the files in his personal safe, they would have been routinely destroyed, and without a record of their prior existence. This is one reason why people who made reports, or were aware of events that took place, don't receive them when they file a FOIA request. Those reports usually, and correctly, have been destroyed.

One of the means to circumvent routine destruction was an old boy net. This did happen with UFOs in some agencies. The preservation was accomplished by having people in key positions being aware of one's interest in the topic and having a copy of reports on the topic forwarded to them. However, that is a hit-or-miss proposition. It is also one that expires when the person leaves the organization. The numerous files that I accumulated while on active duty were destroyed when I retired as there was no one with the personal interest to keep them.

Please note that *Above Top Secret* is not a recognized classification. It is quite reasonable for the general public to be perplexed by the classification systems as they may vary from one agency to another. The labels come in many flavors; however, *Above Top Secret* is not one of them. In UFO circles it is a very popular term, and Mark Allin has even established a Web site by that name. Of course that site is oriented on conspiracy theories and is a great place to catch up on unsubstantiated rumors—some of which may even be true.

At the strategic level the practice of overclassification erodes the public's trust and confidence in their government. Unfortunately, events of the recent past have proven that there are many instances in which officials of the government have been less than truthful. Recent polls have shown that well under half of the American people trust the government. Secrecy, which is justified in some cases, greatly contributes to this declining trust. Also unfortunately, the manner in which the UFO topic has been handled has exacerbated the situation slightly for the general public, but to a great degree among the UFO true believers. Most important, it is not necessary.

Spying

What people generally mean when they discuss a topic as being Above Top Secret is that the material is closely held or compartmented. It is true that one is supposed to have a need to know before being granted access to any classified material. However, there are certain secrets that are more tightly controlled than others. For example, when stealth technology was being developed, that was one of the most closely guarded secrets of its time. Stealth was considered a *war-winner*, or what is now known as a *game changer*. Those are technologies that have such an impact that they can dramatically alter the current rules of warfare. The development of nuclear weapons also was one of those game changers. Given the impact those technologies had, strict access rosters were maintained at all times. Sometimes there are programs that receive such attention. Consider *Ivy Bells*, which was a very expensive and risky program to maintain. In this program American submarines had to operate in the Sea of Okhotsk while U.S. Navy divers installed and retrieved listening devices over submerged Soviet communications cables. Even most of the crew members of the submarines involved were unaware of the real mission they were performing. Unfortunately, a disgruntled and indebted NSA employee, Ronald Pelton, sold the information from the taps to the KGB for a reported $35,000.

Therein lays the need for compartmentalization. Every agency has had employees who became spies for other governments. In the CIA it was Aldrich Ames whose activities resulted in the deaths of several of our assets inside the USSR. The FBI had a senior agent, Robert Hanssen, who also spied for the Soviet Union for twenty-two years. John Walker of the U.S. Navy, along with family members, provided many classified documents to them as well. But the spying was not always for the Soviet Union. Jonathan Pollard, who worked for the Naval Intelligence Service, was caught spying for Israel. His release is periodically raised by Israeli leaders, and vigorously opposed by our Intelligence Community.

There are several common factors regarding the spies who have been caught. First and foremost, they really did hold positions in the Intelligence Community. Their careers were well known and their history could be checked. Contrary to the spy novels, huge sums of money were not involved. While their motivations varied, the degree of financial reward was not a high priority compared to the value of the information they provided.

The Rumor Mill

When we compare the people who have come forward to provide supposedly classified information about UFOs, we find a very different situation. They almost never have documentation that supports their claims. Rather than having official records of their assignments, they often state that other names were used, or their files have been erased. They often hide behind the old adage, *"I'd tell you, but then I'd have to kill you."* Even Bill Coleman employed a version of this saying that led to considerable speculation. He had told selected people that, "We will go for a boat ride. I will whisper in your ear, and then I will come back." People really did ask me to inquire about what message was contained in the whisper, but not catch on to the indication that only one of them would return. Remember, as stated in the beginning, if you want to play, you'd better have a sense of humor.

Bob Lazar is a classic example of an unsubstantiated person revealing unsupported UFO information. Not only do the schools he claimed to have attended not have any records of him, neither do the organizations he stated he had worked for. As proof of government service he produced one Navy receipt for $600. His name did appear in a Los Alamos National Laboratory phone book, but not as an employee. The lab was relatively small and contractors in any capacity were listed along with actual University of California personnel who were employed directly by the lab. A phone book annotation is a far cry from authenticating employment at underground bases in the Nevada desert. Despite the lack of facts, with all of his publicity he laughed all the way to the bank.

In the UFO field, as has been found in military special operations forces, the fraud content by wannabes greatly exceeds the number of real participants. One estimate suggested the ratio of wannabes to real SEALs and Green Berets was about thirty to one. While they can pass in a civilian-dominated bar or at a party, it does not take the real ones long to spot the imposters. There is a language and knowledge base that is extremely difficult to fake for any length of time. In addition, our community is small enough that a quick check will reveal whether or not you are known to the right people.

The same situation is true in the intelligence world. Therefore, when people come forward with extraordinary tales about how they encountered UFO projects in the Black World, they can easily be spotted by those who understand the system and have really worked in it. When former officials like Colonel Bill Coleman of the Air Force or Howell McConnell of NSA discuss their activities in the field, their descriptions fit entirely. They are also careful to differentiate between facts as they know them and speculation. When witnesses to defense-related events, such as Colonel Chuck Halt and Sergeant Jim Penniston at Bentwaters or Lieutenant Bob Salas at the missile silo of Malmstrom AFB, make statements, the records support their claims. However, when there are stories about wars between aliens and humans with mass casualties, underground transportation systems that transit hundreds of miles, and a host of other stories from less credible sources, the facts don't support reality. Unfortunately, the average person has no way to evaluate the accuracy of claims when the cloak of secrecy is invoked.

Technology

There have been times in which advances in technical capability have brought about UFO sightings. Such was the case with the development of the U-2 high-altitude spy plane. At times pilots of commercial airliners would spot a shiny object far above them. To the best of their knowledge, nothing could possibly fly that high. Not knowing they were wrong, they sometimes reported seeing a UFO. For the CIA and the U.S. Air Force, that became an official cover story. It fit their purposes very well and the agencies were very happy to have these inadvertent observations passed off as UFOs.

Today There Are Many More Small Craft Flying

In the current intelligence and combat environment there are a new set of objects that likely will draw such reports. Both the Department of Defense and the Intelligence Community have invested heavily in unmanned aerial vehicles or UAVs. The Air Force prefers to call these aircraft remotely piloted vehicles (RPVs) as there is a pilot involved, not to mention a substantial ground crew necessary to launch the aircraft. They also do not like the use of the word *drones* to describe these functions, as the RPV does not decide where it will go. Rather, the RPV is always being directed and under human control.

There are many new RPVs that are available and the public has become aware of two main systems—the MQ-1 Predator and the MQ-9 Reaper. These are systems that can fly over restricted areas and are armed with missiles while the pilot is sequestered safely up to half a world away at Creech Air Force Base near the nuclear test site in the Nevada desert. Of course missile strikes, especially in Pakistan, have become controversial due to the number of collateral casualties.

When first developed, the Predator was strictly a reconnaissance platform that carried sensor systems over the battlefield. It was tragic incidents like the capture and execution of Navy SEAL Petty Officer First Class Neil Roberts in Afghanistan that brought about the requirement to incorporate weapons. On October 7, 2001, a special operations Chinook helicopter in which he was riding took fire and Roberts fell out. The eyes of Predator, commanders on the ground could only watch in horror as the events unfolded. They could observe his valiant efforts to fight that ended when three mujahideen dragged the wounded SEAL off and shot him.

Since the advent of the Global War on Terror, many new UAVs have entered the arsenal. They range in size from pocket planes that can be launched by hand and see behind the next building to Global Hawk, which can fly 3,000 nautical miles and loiter on station for up to twenty-four hours at altitudes as high as 65,000 feet. These High-Altitude/Long-Endurance UAVs are used to provide coverage of large areas on the ground. Coming soon will be the LEMV or Long Endurance Multi-INT vehicle. A lighter-than-air ship, this craft is being designed to stay on station unmanned for up to three weeks. Flying at an estimated altitude of 20,000 feet, it is very likely that some sightings will mistake this rather large UAV for a UFO.

The United States is not the only country interested in UAV development and production. The military forces of every technologically developed nation have interest in UAV capabilities. Being quiet and hard to spot, they are a favorite tool for gathering intelligence in denied areas. It is likely that these craft will add to UFO sightings. Given the often classified missions, the probability that the users will confirm a UAV is rather low. They will be just another consideration when researching UFO reports.

Minding the Store

In reviewing this manuscript, one of the technically savvy people who has held senior positions in America's space programs took exception to the comment that "no one was minding the store." He noted that, "The Air Force has had a space surveillance system in place for some time prior to SDI that catalogues 9,000 objects in space every day and had 20,000 objects in the space catalogue that are tracked on a monthly basis. The system tracks objects down to thirty centimeters in size by both optical tracking and tracking with a large radar fence. The major purpose of the system is to detect objects in space 'that don't belong there'! In addition, there are many multi-wavelength satellites (from gamma ray, to X-ray to ultraviolet to visible, IR, microwave, HF, and low frequency) that observe the earth 24/7 on a continuous basis, both DoD and spy satellites. These were all operational before 1970. I did write you a paper on this dated August 1995. It is hard to understand how any object in near earth space would not be detected by the US space surveillance system or the dozens of DoD, IC, NASA (LANDSAT), and scientific satellites that have been in operation for some time, all of which are designed to see down to meter to tens of centimeter-size objects."

In fact, he raises some very significant issues and elucidates conundrums that perplex scientists attempting to research UFO sightings. Part of the answer is that some unusual observations have been recorded. There is pretty good evidence that in many instances in which an unexplained event occurs, it is not reported, or at least not forwarded up the chain of command. Young technicians soon learn that making such reports is not in the best interest of enhancing a budding career. In addition, midlevel managers quickly determine that the observation does not constitute a threat, and thus does not need to go to higher levels.

However, another part of the problem is more difficult to rectify. That is when some UFOs appear in hard physical form but do not register on technical sensor systems that should pick them up. Supported by highly credible witnesses, there exist a substantial number of cases in which large UFOs have been seen by people, yet apparently not recorded on sensor systems. There are examples in which UFOs are seen on radar one moment, yet gone the next. In cases such as the Phoenix Lights, the craft was seen and photographed, yet apparently not picked up by Air Force or FAA radars in the vicinity. At a minimum it appears that these enigmatic objects have a capability to manipulate physical reality in manners that we have yet to comprehend. That does not make them less real, just more difficult to explain.

The Bottom Line

Contrary to common belief, there currently does not appear to be any blanket security prohibition against talking about UFOs by members of the military. That was true in the early days. It does seem that overeager security officers did threaten some people, as happened with Milton Torres. I can state that in my personal experience, being very vocal about discussing UFOs, no one has ever officially warned me about secrecy. When I have talked with people who have observed UFOs, many believe that the prohibition exists, yet none ever had an official debriefing. I have a standing offer to any credible person who has a story to tell. If they want clarification about official classification of specific incidents in which they were personally involved, then I will assist them in getting the information out. *It is safe to talk!*

Summary

The U.S. Government is guilty of extensively overclassifying information. Some of that is to hide stupidity and blunders from public view, but more frequently it is simply because the officials don't know what to do with the data. Most importantly, regarding UFOs, the preponderance of evidence suggests that any classification that was done occurred when relatively low-level investigators invoked secrecy in an ad hoc fashion, rather than it being a formalized policy issued by a duly authorized agency. The most striking case of such behavior was that of the 1980 Bentwaters incident. While some enlisted men were questioned and sworn to secrecy, Colonel Halt received no such admonition.

To this day there are no follow-up procedures in place by which individuals are notified when secrecy has been lifted on a specific topic. The importance of that was demonstrated by the emotional burden that some unidentified agent placed on Milton Torres decades ago. Torres honored his agreement, even when it meant his father would die without learning of his son's dramatic encounter with a craft that was certainly not of earthly origin.

Except for discussion of sources and methods, there is no indication that UFOs hold any systemic classification. For those military personnel who have spoken out—and there are many—none have ever been sanctioned for revealing officially classified information.

In addition to U.S. classification of UFOs, much has been made of our NATO allies. Discussed next is the stories about their sightings and reports.

WHAT NATO KNOWS

The day will come undoubtedly when the phenomenon will be observed
with technological means of detection and collection that won't
leave a single doubt about its origin. . . . But it exists, it is real,
and that in itself is an important conclusion.
—MAJOR GENERAL WILFRED DE BROUWER, DEPUTY
CHIEF OF ROYAL BELGIAN AIR FORCE

It has been reported that the North Atlantic Treaty Organization (NATO) has special knowledge about UFOs. A good deal of this information comes from a retired U.S. Army Command Sergeant Major Robert O. Dean who was assigned at Paris in 1964 at what was then SHAPE Headquarters. SHAPE was the Supreme Headquarters Allied Powers Europe, and was later moved to Mons Belgium. SHAPE was an element of NATO and thus the information provided does fit the organizational scenario. For the record I have met Command Sergeant Major Dean on several occasions and found him to be an affable gentleman. However, we agree to disagree about NATO and the UFO topic.

An Alleged NATO Document

Command Sergeant Major Dean states that while assigned at SHAPE he was given a document to read called *"The Assessment,"* which allegedly dealt with UFOs. According to Dean the document was classified Cosmic Top Secret. While some researchers question the existence of that classification, it is confirmed that it is real and appropriate for the position he held at the time. In fact, most officers and senior noncommissioned officers assigned in NATO would have held that clearance. The total number of people holding a Cosmic Top Secret clearance over the decades would be in the tens of thousands. In fact I still hold that level of clearance even though I have never been assigned to a NATO organization. However, I have participated in several NATO studies that allowed me to explore the UFO topic a bit further.

Dean states that this document was very limited, with only fifteen copies being prepared. Exactly why he was allowed access to this sensitive information remains unclear. His description was more like it came from his classified reading room. He also claims that the document confirmed that at least four extraterrestrial civilizations had visited Earth by 1963, but by some accounts the number had increased to twelve by 1976. One must assume that Earth got noticed and put on the vacation list by some galactic travel agency. However, in recent years Dean's accounts of what was happening have taken the ET visitation concept quite a ways further. He has reported that since some of the ETs looked totally human they had extensive contact on Earth and he believed he had met some of them. In interviews he noted that some of the admirals and generals couldn't deal with the notion "that these guys [meaning ETs] could be walking up and down the corridors of SHAPE headquarters or the Pentagon or the White House." In fact, Dean believes that they have made such visits.

His accounts of interaction between military forces and the extraterrestrials include numerous violent confrontations. According to Command Sergeant Major Dean, an order was given by an unnamed general officer to engage the UFOs. This, he reports, started a one-sided shooting war using guns and missiles. The conflict allegedly lasted only ninety days and in that time Dean contends that we lost thirty aircraft, though most of the pilots were able to bail out of their disabled planes. It is hard to believe that losses of that magnitude in a short period, except during open warfare, would be kept from the public record for so long. This would have caused an uproar within the Air Force. By comparison, during the little acknowledged shooting incidents between the United States and the USSR, sixteen aircraft were lost over about a decade. Those cases initially were reported as accidents, but the real cause came to light during Congressional hearings in 1992.

As reported in the previous chapter, there were attempts made under orders to shoot at UFOs when they intruded on sensitive airspace. But as Lieutenant Torres found out in the United Kingdom, they could easily outdistance any conventional aircraft of the time. In fairness to Dean, there were some other attempts to intercept UFOs that ended tragically. The case of Captain Thomas Mantell crashing after chasing such a craft has fueled that notion for decades. On January 7, 1948, flying a P-51 Mustang for the 165th Fighter Squadron of the Kentucky Air National Guard, Captain Mantell was advised to intercept an object that had been under observation by multiple witnesses for some time. It had been seen close to the ground, and then it rapidly ascended to an estimated 10,000 feet. Others in the pursuit were Lieutenant Albert

Clemmons and Lieutenant Hammond, each piloting his own plane. As they approached the UFO continued to climb. At 22,500 feet Lieutenant Clemmons and Lieutenant Hammond broke off pursuit as they were low on oxygen. Captain Mantell continued to at least 25,000 feet whereupon he apparently lost consciousness and went into a tailspin all the way back to the ground. It was noted that his watch stopped at 3:18 P.M., which was the time of impact. There is no doubt that Mantell crashed, and that it was related to a UFO sighting. The conflict comes from whether or not he simply passed out from lack of oxygen at high altitude, or as some researchers claim, his aircraft was destroyed because it got too close to the UFO. The case has never been fully resolved.

Obviously there are serious problems with the story of Command Sergeant Major Dean. His narrative would mean that the existence of UFOs and alien visitors would have been known by many of the senior officers at SHAPE. As a NATO element, those officers came from countries throughout Europe, plus Canada and the United States. Such knowledge would have spread like wildfire throughout the continent, but there is no record of that. In addition, the information would have made it to the Soviet Union. The Soviet KGB was very effective in penetrating NATO and had copies of the most sensitive files it had. The visitation of extraterrestrial civilizations is not something the KGB would have missed.

Discussions with Senior Leadership

My experience regarding UFO information and NATO came about in the mid-1990s. Based on my work with nonlethal weapons I was appointed to be a representative of the United States on several studies about their potential. Also, in 1993 I attended the Senior Executives Course on National and International Security at the John F. Kennedy School of Government at Harvard University. This is a program attended mostly by military flag grade officers. However, they also allow a few civilians of comparable grade to participate. Then working at Los Alamos National Laboratory as a project manager, I was selected to attend the course. During this program there were daily small group discussion sessions at the beginning of each day. About ten or twelve people would meet in the same group, which was headed by the most senior person in attendance. In my session, that person was U.S. Navy Vice Admiral Norman Ray.

Following that course at Harvard, Vice Admiral Ray was assigned as the NATO Assistant Secretary General for Defense Support at the headquarters in Brussels, Belgium. This was the number two military position in NATO and it was always held by an American because of control of nuclear weapons. Shortly thereafter I was involved in another study under a research organization called AGARD, located in Paris. While the first meeting of the studies was held in the French capital, subsequent sessions were conducted at various locations in Europe. One of the more lengthy writing sessions was held in Brussels so I took advantage of the location to arrange a courtesy call on Vice Admiral Ray. We talked about a range of topics, and this should not be viewed as a UFO meeting per se. However, since he knew me, I took the opportunity to raise the issue with him. Given his position, if there were any sizable effort in the field, or UFOs were viewed with serious concern, he would have been in a position to know. His response was quite similar to what I usually heard when briefing other senior officers while running the Advanced Theoretical Physics project. He was not evasive, but indicated that he was not aware of any efforts in the field.

Over the year we spent on the study I had the opportunity to raise the same question with other European officers who were generally lieutenant colonels and colonels. None had specific knowledge of the topic, but several were quite curious, especially the Belgians. They, after all, had experienced a substantial number of sightings over their territory.

The Well-documented Case with Then Lieutenant Colonel De Brouwer

In 1989 and 1990 there were repeated sightings of a large craft over the Belgian countryside. Evidence included multiple radar contacts, photographs, and eyewitness reports from thousands of people. The most famous case took place the night of March 30 to 31, 1990, and involved pursuit by Belgian F-16s. In addition to the multisensory data, the flight characteristics of this event defy what is known about conventional aircraft. At about 11:00 P.M. a gendarmerie called the Control Reporting Station at Glons and reported a stationary object that was illuminated with three lights that were changing colors from red to green to yellow. They were quite distinguishable from stars. Then a radar station reported contacting an unknown object that was moving west at about 25 knots.

After observing some erratic moves, two F-16s were launched just after midnight and were vectored to the area of the sighting. The pilots had brief radar contact and even locked on to the target. The UFO was extremely uncooperative, changing speed from 150 knots to 970 knots in a very short period of time. It also quickly dropped in altitude and the pursuit aircraft were notified they were now above the UFO. The Glons radar lost contact with the target, but another system picked it up a few minutes later. That radar system appeared to have been jammed by the UFO. The cat and mouse game continued for about one and a half hours before multiple UFOs departed in four different directions. Ground observers saw both the F-16s and the UFOs, but the pilots never had visual contact.

Part of what makes this remarkable incident so important was the Colonel who made the report was highly credible. Colonel Wilfried De Brouwer would go on to become a major general and the Deputy Chief of the Royal Belgian Air Force. The official report stated, "The aircraft had brief radar contacts on several occasions" but when "the pilots were able to secure a lock on one of the targets for a few seconds . . . a drastic change in the behavior of the UFOs [occurred]." During one of these locks, "The speed of the target changed [quickly] from 150 to 970 knots and from 9,000 to 5,000 feet [altitude], returning . . . to 11,000 feet [changing] again to close to ground level." De Brouwer explained that this extreme acceleration, equivalent to 40 gs, would exclude a human pilot. In tests it has been shown that some people blackout with as little as 2 g-forces. With extensive training a few fighter pilots have withstood 9 gs, but not beyond. Obviously, 40 g acceleration, or anything close to that, would eliminate any known human-operated system.

Still, the initial speculation was that the UFOs were some sort of secret American project. The Belgian Government officially asked the United States for a response. The Belgian guesses about the nature of the craft ran from the F-117 or B-2 to some sort of unpiloted aircraft. At our meetings a few years later, the officers asked me the same questions. Clearly, these intrusions in NATO airspace are among some of the most solid cases that have been recorded. Later I did have the opportunity to talk with General De Brouwer about the experience. He has made several public appearances and stands by his evidence. Now in retirement, he has concentrated his efforts in humanitarian relief work around the world.

Canadian Accusations Get Crazy

Canada is a NATO member and due to proximity and historical interests enjoys a special relationship with the United States. Therefore, it is disconcerting when former members of their most senior leadership engage in provocative and unsubstantiated comments inferring contact with ETs and far worse. On multiple occasions I have heard the Honorable Paul Hellyer, former Minister of Defense of Canada, and longtime member of their Parliament, inform audiences that the United States has been involved in cooperative activities with extraterrestrial civilizations.

On April 30, 2008, at the *Exopolitics* conference in Gaithersburg, Maryland, he made the following quite unequivocal statements. "Decades ago visitors from other planets warned us about where we were headed and offered to help. We [unidentified government officials] interpreted their visits as a threat and decided to shoot first and ask questions after." As with reports from Command Sergeant Major Dean, he states that some of our planes were lost during these armed encounters. According to Hellyer, "Military aircraft are built with a small margin of safety." His thesis suggests that therefore the materials are more likely to disintegrate when the aircraft get close to UFOs. Remember, these comments are coming from a former Minister of Defense concerning the safety of our military aircraft. They are frankly ill-founded.

Rather aggressively he went on to state that *"the U.S. military wanted to use nuclear weapons [against the ETs] but their use would result in annihilation of us all."* Hellyer said, *"The military is so paranoid they use the visitor's technology to fight them off, rather than welcoming them as partners in development."* Here he is confirming that not only do we have advanced alien technology, but we have weaponized it as well. He stated that the United States *"has developed UFOs that are indistinguishable from the visitors."* As a side comment, why then have we not seen these advanced weapons on the battlefields of Iraq and Afghanistan?

Hellyer flatly states that the United States has spent trillions of dollars on *"programs about which the President and Congress have deliberately been kept in the dark."* It is from this black budget that the UFO program has been developed. According to his thesis, someone has developed free energy but did not tell the Secretary of Defense or the President "because they don't have a need to know." This premise is absurd.

In fairness, he did take the statement about unaccounted for money from remarks of then Secretary of Defense Donald Rumsfeld. This was just before 9/11 and Rumsfeld was about to take on the large defense contractors and reform business practices. There is no doubt that accountability has been a weak spot in American defense spending, and it was probably worse immediately following the initiation of GWOT. Still poor accountability does not mean that Congress has no oversight of all programs. To argue that a modest-sized program could have been executed from funds siphoned off illegally is a fair conjecture. To suggest that trillions have been diverted into UFO technology development seems far-fetched. He should also know that the exclusion of the Secretary of Defense and the President from information of that magnitude based on lack of need to know is ridiculous.

Hellyer went on to raise another established figure in Canadian Government,

Wilbert Smith, who he believes had direct contacts with ETs. Hellyer stated, "Wilbert Smith asked the visitors about the accidental destruction of aircraft flying in the vicinity of flying saucers." The ETs allegedly responded that "They had now taken corrective means to avoid our aircraft." Smith was an employee of the Canadian Department of Transportation, and as such wrote an internal memo that has received much attention in UFO circles. Smith wrote the memo in 1950, and the world at that time was very different. Concern about UFOs was far higher than it is today.

Hellyer's personal experience in the UFO field is, by his own admission, very limited. Mostly it is based on the Corso book and a Peter Jennings television special. He stated that some undisclosed American general officer had told him "to believe every word of Corso." At the Society for Scientific Exploration conference in 2009 I asked a question during his presentation. Specifically I asked if he believed that the Cold War was a cover for fighting ET as Corso had claimed. Hellyer's response was that he "was not going to nitpick!" My retort was that I did not believe that subverting the rationale for the Cold War was a nitpick. Privately I asked Hellyer about the identity of the general who had told him to believe Corso. His response was to ask me how I "ever got to be a colonel asking a minister about his sources." We later sat on a panel together (with people in between us), but I doubt that will happen again.

What is most disconcerting is that a senior official of a close ally could have such an inadequate understanding of how our government functions. This was a former Canadian Minister of Defense telling audiences that the United States and its allies have engaged in an undisclosed shooting war with ET.

The French Connection

France is one of the most influential members of NATO and has a long-standing official interest in UFOs. In fact, it established a small office in the French equivalent of NASA, the National Center for Space Studies (Centre National D'études Spatiales) or CNES. That office, GEPAN, was created in the mid-1970s in large part due to the intense wave of high-strangeness UFO sightings in France in 1954. Over the years GEPAN has done some fairly intensive studies of cases that provided hard evidence that the events were real.

In 1999 there was a UFO report published in France that attracted considerable attention worldwide. *The COMETA Report,* as the document became known, was given the trappings of an official study, when in fact it was conducted by members associated with the Institute of Higher National Defence Studies, or IHEDN (Institut des Hautes Études de Défense Nationale), a private think tank. The composition of the study panel provided considerable, but probably unwarranted, credibility. The panel was headed by General Denis Letty of the French Air Force, and membership included General Bruno Lemoine of the French Air Force; Admiral Marc Merlo; Michel Algrin, doctor in Political Sciences and attorney at law; General Pierre Bescond, engineer for armaments; Denis Blancher, Chief National Police Superintendent at the Ministry of the Interior; Christian Marchal, chief engineer of the national Corps des Mines and research director at the National Office of Aeronautical Research (ONERA); and General Alain Orszag, Ph.D., in physics and engineer for armaments.

Once completed, the report was sent to General Bernard Norlain of the French Air Force, former Director of IHEDN, who wrote a preface. Also a preamble prepared by André Lebeau, former President of the National Center for Space Studies, was included. Unfortunately they did not see the final portion of the study. Added after the document was forwarded for comments, that section was decidedly anti-American and accuses the United States of a major cover-up.

In the end, the panel accepted Roswell as a real event and believes the United States has been hiding the information. They claim the Americans have proof of a threat, and the rationale they provided for U.S. continued secrecy includes the usual suspects:

— Potential for social upheaval
— Panic
— Religious fundamentalism

They concluded that the Pentagon controls this knowledge for "the United States to hold a position of great supremacy over terrestrial adversaries while giving it considerable response capability against a possible threat from space."

Strikingly, the bibliography of the study contains a long list of popular books that would be well known to those who read the UFO literature. A concern should be that this panel appears to have given scientific credibility to those publications.

The panel also manipulated the release of their report to make it appear as if it were an official French Government study. One of the tricks they employed was to send the report to the President, while making it appear as if it were requested officially. Note that this is also frequently done by some U.S. provocateurs. In fact, the COMETA Report was published in a tabloid, and proved to be an embarrassment to IHEDN, which is a highly respectable organization.

In June 2010 Mario Borghezio of Italy introduced a measure to the European Union Parliament, which includes most of the NATO nations, requesting release of classified UFO information by all member nations. The document cites a UN resolution and states, "whereas in 1978 the 33rd General Assembly of the United Nations formally recognized UFOs as a valid issue." In fact, on July 14, 1978, Secretary General Kurt Waldheim did meet with J. Allen Hynek, Jacques Vallee, astronaut Gordon Cooper, and others regarding UFOs. On November 27 of that year a news conference on the topic was held at the UN Headquarters in New York.

That citation alone suggests the relative importance that the UN and its member nations afford this issue. Rightly or wrongly, with more than three decades intervening, not much official interest in UFOs has been shown by any significant body of governance. Despite the considerable time and effort necessary to generate and execute these high-level meetings, the nearly imperceptible response is indicative of their general lack of effectiveness from a practical perspective. As with congressional hearings, unless there is a plan in place that is both politically supported and adequately resourced, conducting coruscant engagements has minimal pragmatic impact.

Summary

As the most extensive defense organization in the world, NATO crosses the intelligence services of many countries, and has a centralized reporting system. There is a great disparity in the quality of information available regarding UFOs, and some of it seems frankly preposterous. However, there are observations, such as those made by General De Brouwer, that undeniably support the fact that unidentified craft were operating over Europe. His report was substantiated by multiple technical sources and credible photographs of UFOs are available.

What is of serious concern is that the IHEDN panel, consisting of several high-ranking officers with extensive experience in both NATO operations and bilateral agreements, willingly accepted information from popular sources and without any confirmation acted upon it. It should be extremely disconcerting that we have both French and Canadian senior officials that distrust the United States as an ally, and are willing to believe, and even act upon, information that is so poorly derived, counterintuitive, and has not been validated by credible sources.

It is extremely important to note that senior officials from several governments were willing to make serious judgments about both the nature of UFOs, and the involvement of the U.S. Government based on unsubstantiated data from questionable sources. The implications of such poorly conceived conjecture in the realm of international policies and relationships are horrendous and could have disastrous consequences.

While there is no doubt that some UFO sightings constitute real physical objects that display remarkable aeronautical characteristics, the range of observations suggests that there is no single simple answer. The next chapter resolves several mysteries while exposing many others.

CONSIDERATIONS, SPECULATION, AND PUZZLES ADDRESSED

There are more things in heaven and earth, Horatio,
than are dreamt of in your philosophy.
—WILLIAM SHAKESPEARE, *HAMLET*

Considerations

One of the most dangerous positions one can take in an area as complex as UFOs is to be dogmatic about convictions. This is a fatal error, and one made by many UFO researchers who have written on this topic. Over the past few decades I have given a great amount of thought to the various possibilities that might explain the facts. At the end of the day I come down relying on the firsthand experiences that have been gained in this exploration. As stated earlier, good people can look at facts and come to different conclusions.

Agree to Disagree

It is important to note that I have several friends whose experience and opinions I highly value, but who have come to different conclusions than I have. This is particularly true when it comes to whether or not material from a UFO crash ever came into the possession of the U.S. Government. I have concluded that it did not and I'll discuss Roswell in more detail shortly.

Some of the people involved are known to the public as interested in UFOs, but others prefer to have their identity protected—and that will be respected as it is not relevant to the discussion. To the best of my knowledge, each of those individuals has based his or her opinion on secondhand information from a trusted source. In some cases their sources are known, but in others they are not. None of these people with whom I respectfully agree to disagree believe that the U.S. Government has successfully reverse engineered a UFO.

The alternative position that is most frequently offered is that a crash did occur and recovery of some material made. The understanding is that some high-level scientific body was convened to examine the evidence and advise the government as to what we could learn from the technology that was believed to be far more advanced than our own. The consensus of the alternative position would be that whatever was examined was beyond the comprehension of those scientists. Thus, they decided to sequester the material in a safe location and arrange for periodic reviews to determine whether or not progress could be made in understanding the physics associated with that UFO. As I indicated at the beginning of this book, this position is very close to what I initially believed to be true.

As mentioned earlier, an analogy provided regarding a crash was to consider what would happen if an advanced airplane like the F-22 Raptor, built for stealthy air superiority, plummeted into the Amazon jungle and was found by some of the uncontacted people who still inhabit that vast and remote area. While the tribespeople may have an excellent understanding of how to survive in that harsh environment, they would not be able to take advantage of any of the advanced features of the aircraft. They may have seen airplanes fly over, but the fundamentals of aviation would still elude them. The basic concepts of propulsion, avionics, communications, or weaponry would not be understood, let alone how the plane's signal suppression technology allowed it to evade sensors the tribe has no idea exist. The most likely advantage that they would gain would be to craft better spears and arrows from shards of material scattered about the crash site. If the pilot was killed in the crash it is even probable that he would be eaten, not examined for anthropological value. As the UFO analogy goes, we are those uncontacted tribespeople.

One of the harder disconnects to rectify comes from a very senior NASA scientist. Having discussed UFOs several times, he firmly believes that there is a black program somewhere. Of course, that would not be in his organization. Like all of my other friends with whom there is agreement to disagree, the information is secondhand and whoever *THEY* might be are not identified. Interestingly, he also has technical problems conceiving of long duration, deep spaceflights for humans due to the excessive radiation they would encounter.

All of the people referred to as trusted colleagues are aware of the ATP study, and some of them were involved in it. Still, none of them has had the experiences that I did in briefing upper-echelon executives. One argument is that I just didn't push the right buttons, missed the organization(s) responsible for such a program, or was simply intentionally misled by those who were cognizant of the program.

Of course, any or all of these situations are possible. I tend to discount the argument that I was intentionally misled. When conducting the briefings and interviews, the reaction of each key person was watched closely and their behavior evaluated. Not once was intentional deception or deflection detected. More frequently the reaction was more like a deer being caught in the headlights as the topic for discussion was often not conveyed to the executive until I began speaking. In addition, their behavior, which often included providing assistance, was counterintuitive to protecting a hidden program. As mentioned, I have run into compartmented programs on other topics and can tell immediately when a wall has been hit. Never did that happen with ATP briefings. Could I have missed it? Possible, but unlikely.

In addition, I have cross-referenced my experiences with all of the logic sequences that have been described in this book. From the Apollo program to energy issues, national security, and a host of other factors, the evidence is highly consistent that the U.S. Government is not actively involved with UFOs, even when incidents have suggested they should be. It must be admitted that the pieces are not all complete, but the vast majority of them are in place, and the picture pretty clear.

The Skeptics

"After more than five decades filled with UFO reports, books, TV programs, magazines and conferences, what is the case for UFOs?" That was the printed description for a presentation by Robert Sheaffer at the 2004 national convention of Mensa, the high-IQ society. Sheaffer, who describes himself as a skeptical investigator of all manner of bogus claims, is a vocal member of the Committee for Skeptical Inquiry (formerly CSICOPS), who often appears as a spokesperson for the organization. Held in Las Vegas that year, I also gave two presentations at the Mensa conference including one on the Global War on Terror and another on UFOs. The opening slide included Sheaffer's statement and added, "The evidence is clear and overwhelming, and here it is!" Sheaffer and Ken Frazier, the editor of the *Skeptical Inquirer*, were sitting right in the middle, just a few rows from the front. While they listened politely and did not make any attempts at questions that could embarrass me, it is obvious from their later writings I didn't make a dent.

Nor is it likely that any amount of evidence would persuade them to change their minds. While I consider myself to be rather skeptical of many claims, with good reason, they would better be classified as debunkers. The basic premise is that these anomalous events can't be true, therefore they aren't. If they cannot disprove the anomaly, they assume that the prosaic, commonplace answer has not yet been found—but they are equally confident that one exists. Facts are not of consequence, just inconvenient details and often ignored.

That is another prevalent characteristic with skeptics—a proclivity for intellectual arrogance. In *Future War* I addressed this problem as *cerebralcentrism,* a typically American position in which we believe that we are smarter than everyone else. As an example, in 1980 while attending Command and General Staff College at Fort Leavenworth, Kansas, a classmate who was unofficially assigned as the first U.S. Army astronaut, stated quite confidently that 99 percent of everything we could learn from science had been already discovered. According to this engineer, all that scientists are then doing was tidying up the loose ends. Similar intellectual snobbery is epidemic with many skeptics.

In the case of UFOs, flares and astronomical bodies are favorite explanations. And, in truth, sometimes they are responsible for misidentification. But when you obtain the kind of evidence described in Chapter 10, the skeptics' efforts at conventional explanations fail the test of common sense. Seriously, when Jim Penniston talks about the UFO he touched, yet skeptics attribute the sighting to a lighthouse that is miles away, then there is a precarious lack of logic involved. This also applies to the Phoenix Lights in which flares over a range at the wrong time are still used to explain a massive object floating over the city in the early evening and seen by thousands of people as it obscures the stars.

In the face of overwhelming physical evidence, supported by credible eyewitness testimony, the skeptics' astronomer, James McGaha, still gets away with simply claiming that UFOs are a *"conspiracy wrapped in suspicion and myth."* This is reinforced by a *Fox News* commentator asking, *"If ETs are so highly evolved why do they only contact people named Bubba?"* And then in 2009 Ken Frazier wrote, *"Unfortunately, UFO reports seldom if ever come from anyone really knowledgeable in trained observation of the skies and unusual phenomena."* He goes on to explain that law enforcement officers and pilots do not meet his minimum qualifications.

To accept those comments we must assume that Colonel De Brouwer was unqualified to make his radar observations about a UFO he pursued over Belgium; that neither Lieutenant Torres nor Lieutenant Jafari were qualified when they prepared to engage huge UFOs with weapons; that NORAD personnel were unqualified when reporting fastwalkers; or that the ICBMs did not fail; and that Soviet pilots did not encounter UFOs. Ah yes, unqualified Bubbas all! Torres went on to get his Ph.D. in engineering and become a professor, De Brouwer became the Deputy Chief of Staff of the Royal Belgian Air Force, Jafari became a lieutenant general in the Iranian Air Force, and Tretyak was the Russian Deputy Minister of Defense. Then too Arizona must have elected Bubba to be their governor.

In truly anomalous cases, the fundamental approach of skeptics is to find some simple flaw in the report, then leap to the conclusion that if a detail is questionable, the entire observation must be discarded. Contrary to their protestations, these sightings rarely have single point failure. In general debates, the skeptics/debunkers rely on discussing weak cases and then extrapolating to the whole field. This is a false position as UFO proponents already agree that around 95 percent of reports have an ordinary explanation. There is also the premise that if someone can fake it, then so did the person making the report. This is particularly true with photographic cases. Frankly, with the advances in digital photography, pictures can only be used to support other data as it is almost impossible to detect fraud.

Unfortunately, there is considerable fraud in the field of UFOs and related matters. Many people believe it is funny to fool the public, and will at times endanger others to accomplish their goals. This was the case in Morristown, New Jersey, when two boys sent balloons aloft carrying burning road flares and claimed that it was a social experiment. The joke took precedence over the fire hazard. What is far worse, the culprits were actively encouraged by the skeptics as can been seen in the April 19, 2009, edition of *eSkeptic*. The perpetrators were prosecuted, but unfortunately only sentenced to community service. They seem to be unaware that the Japanese had attacked America during World War II using basically the same technique. We will never see an end to hoaxing, but the consequences are far from benign and certainly provide constant fodder for the skeptics.

The Skeptics Have an Atheistic Component

Interestingly, skeptics have many of the characteristics of true believers. The exception is that in the case of UFOs they truly believe in the impossibility of these events, or whatever else they're against. Also, as Hal Puthoff notes, the skeptics are not skeptical enough! They easily are skeptical of what one should be skeptical of—the reported event. However, they are not sufficiently skeptical to question the views—ones that often do not fit the facts of the situation—that are aligned with their own mythology. As Jon Stewart, host of *The Daily Show* on *Comedy Central,* correctly stated, *"Irrational thinking will never be overcome by empirical evidence."*

There is an underlying primordial issue with most skeptics/debunkers. That is fear. At risk are the foundations of their belief system. In attending their conventions, you will find that *atheism* is quite a common theme. The manifestation of those thoughts runs from a simple personal belief to overt hostility toward religion. This animus is best personified by self-avowed skeptic Penn Jillette, of Penn and Teller, who proudly displayed his anti-Christian T-shirt at *The Amazing Meeting* sponsored by James Randi here in Las Vegas. A mecca for skeptics, more information about those meetings can be found on the *Atheist News and Views* Web site.

In *The Warrior's Edge* I wrote about a scientist who built a special instrument for me. He was very concerned about how it would function and bluntly stated, "Don't tell me something that says I must relearn physics—because I don't want to hear it." At least he was honest about his motivation. Most of the skeptics are not. If the characteristics displayed by UFOs and other phenomena are real then their worldview is in jeopardy and they don't want to hear about it.

Speculation

Precognitive Sentient Phenomena

There is a concept I propose for consideration termed *precognitive sentient phenomena* (PSP). This is derived from multiphenomena observations and it offers a partial explanation for UFOs and other unusual sightings. It is a mistake to look at phenomena through academically imposed stovepipes. There is an attempt by most scientists to immediately delimit what is being researched. This is a normal, and often proper, approach. Obviously, researchers need to clearly state the parameters of their exploration or the topics become so broad as to make little sense. The problem in studying phenomena is that most researchers don't understand the boundaries, and thus frequently exclude issues that should have been brought inside the study. It is a Catch-22, but one that needs reexamination.

The issue of *The Trickster* is well established in paranormal research. That means that whatever is generating the incidents does so in a manner that does not remain consistent over time. What is being proposed is a derivation of that idea. The precognitive sentient phenomena concept suggests that there is some external controlling agent that initiates these events that are observed and reported. It appears as though that agent not only determines all factors of the event, but is already (i.e., precognitively) aware of how the observers or researchers will respond to any given stimuli. The agent can be considered like the Trickster that is always in control of the observations. Every time researchers get close to an understanding of the situation, the parameters are altered or new variables are entered into the equation.

Phenomena Are Numerous—and That Is Problematic

Among the phenomena observed around the world are many types of UFOs ranging from orbs or balls of light floating in the air to large, hard crafts that are registered on sensor systems and widely photographed. However, it is this diversity of observations that has perplexed anyone attempting to research the field. Rather than calling the observations UFOs, unidentified aerial phenomena may better describe the full variability of sightings.

It was just such discrepancies that caused Nobel Prize–winning physicist Richard Feynman to denigrate the field. Eccentric, Feynman was well known at Los Alamos for his idiosyncrasies that included late-night forays into deserted and locked offices that contained some of the nation's most precious secrets. There he would open safes for which he did not have the combination and leave a note inside for the unsuspecting scientist. Having been at Los Alamos from the Manhattan Project days, and closely associated with many of the best nuclear scientists in the world including Hans Bethe, Niels Bohr, and LANL director Robert Oppenheimer, Feynman's disdain for UFOs is yet another mark against the notion of recovered material from Roswell.

However, I'll go so far as to suggest that Feynman's argument against UFOs was wrong, as it was based on a limited understanding of the full parameters of the subject he was addressing. He did appear to base that opinion on the extraterrestrial hypothesis that there were visitors from distant planets observing Earth and interacting with us on occasion. To accept the ET hypothesis generally infers that the ETs must get here from there (wherever that may be). If bound by the speed of light as the upper limit for travel, the distances involved present obvious problems. Of course, most readers are well aware of what *Star Trek* termed warp speed, and there may be means by which superluminal travel is possible. In fact, there are scientists working on those problems today. Even so, observations spanning the realm from vaporous orbs to hard craft the size of aircraft carriers—occasionally larger—make it difficult to support the relatively simple ET hypothesis as the origin for some, or all, of the truly unexplained UFO reports. By excluding external data from consideration, the variety of observations does suggest that there are just too many different types of UFOs reported. Most scientists have a hard time accepting the possibility that one civilization might have arrived, while the prospect of a multitude doing so seems unfathomable.

The Extraterrestrial Hypothesis Is Too Limited

What is suggested is that there is a need to cast a broader net for content and not just limit the input to sightings of hard physical objects. Other researchers have made similar observations including Jacques Vallee in several of his books. We agree that at times hard, physical craft exist, but there is much more to the phenomena. Unfortunately the true boundaries are unknown. It seems certain that human consciousness is at least part of the formula. As Jacques and I have discussed privately, the observations of UFOs are frequently just camouflage. In *Messengers of Deception* Jacques stated, "I propose: that the UFO we see is a device which creates a distortion of the witnesses' reality; that it does so for a purpose, which is to project images or fabricated scenes designed to change our belief systems; and that the technology we observe is only the incidental support for a worldwide enterprise of 'subliminal seduction.'" Readers can find much more of Jacques Vallee's thinking in *Passport to Magonia* or his later trilogy *Dimensions, Confrontations,* and *Revelations.*

It is striking to note how the reports have changed over time. In 1896 and 1897 huge airships were reported in the western United States. Their characteristics make no sense from our current understanding of interplanetary travel or even long-distance earth transport. But at the time they raised quite a stir and the witnesses numbered in the thousands. Also noteworthy is a case that Jacques investigated dating from the early 1900s. A driver in a rural area rounded a hill and was confronted by a shiny hexagon-shaped craft with windows hovering low above the ground. That driver reported that the details of the craft were so clear that he "could see every rivet." Such an observation is very significant. At the time of the sighting the object would have been taken for advanced technology. Yet any contemporary aeronautical engineer would know that rivets in aviation were relatively short-lived and certainly not used in the exterior of a spacecraft. This observation points to the conclusion that many of the physical craft that are seen tend to be in advance of current technology, but not beyond the understanding of human consciousness. In other words, they are a few years, or possibly decades, ahead of extant capability, but not the ultra-intelligent leaps that are often attributed to them. That means the commonly used phrase "that they are a thousand years ahead of us" is wrong. It is therefore my contention that if there is any hope of understanding the phenomena, it is imperative that the search be expanded, not delimited.

This discussion suggests Feynman came to the wrong conclusion because he made limiting assumptions. At no point did he allow for the role of consciousness in the understanding of UFOs. It should also be noted that Feynman made similar disparaging remarks about parapsychology and seemed to believe that scientists who believed in any psychic phenomena had committed the mortal sin of fooling themselves.

Enter Stephen Hawking

In the spring of 2010 British physicist Stephen Hawking dramatically entered the discussion concerning the probability of alien life elsewhere in the universe. Wheelchair-bound, Hawking is renowned for his accomplishments in theoretical physics while experiencing the physically devastating symptoms of ALS. In his *Discovery Channel* series, *Into the Universe with Stephen Hawking*, he proved prone to say far more than he knew, and he was not acquainted with some of the basic facts about UFO phenomena.

Unfortunately, it was his dire predictions about the interests aliens might have in Earth that caught the attention of major media outlets around the world. The basic premises of the intentionally alarming headlines were that we should *beware of aliens as they may not come in peace*. Possibly existing as intellectually and technologically advanced nomads, Hawking hypothesized these scavengers might have depleted their home planet and are now traversing the universe in search of life-sustaining resources. Their purpose, he contended, would be to conquer or colonize inferior species, taking what they need and not sharing their knowledge. While this is the common fodder of Hollywood fantasies, it now was being espoused by a respected scientist. Hawking even noted that humans should not be sending invitations to these unknown entities, just in case they are hostile. While he is not the first to so comment, with this simplistic pronouncement he appears to ignore the fact that humans have literally been broadcasting our existence over the radio frequency spectrum for more than a century. It is far too late to recall the plethora of signals that are permeating the universe at an ever-expanding rate. The secret is out for anybody who might be listening.

Hawking's analysis was the product of his "mathematical brain," albeit bounded by constrained assumptions. He noted that the "laws of life" would be universal, just as he believes the laws of physics are. The problem with that statement is that we clearly do not know the limits of life here on Earth, let alone the rest of the universe. For example, for many decades it was assumed that photosynthesis from the sun was an essential ingredient for any form of life. It was only after Robert Ballard and his research team discovered living organisms 8,000 feet below sea level, in an environment that had been believed to be too inhospitable to support life, that biologists began to comprehend how that could occur. No one had dreamed that black smokers would exist in a poisonous milieu at temperatures as hot as 300 Celsius (572 Fahrenheit) and convert toxic chemicals into nutrition for the newly discovered life-forms. Given the relatively recent discovery of the chemosynthesis process, it is therefore reasonable to question how much is known about the full parameters under which life could exist.

Hawking and I both agree that from a purely mathematical perspective there is a high probability that life does exist elsewhere. Just the number of planets that should exist in what are sometimes called the *Goldilocks regions* would predict a positive response. That says nothing about the form such life might take. However, there is also the *Rare Earth Theory* that suggests that the template for life is so narrow, and that the ingredients and circumstances necessary so unique, that life on Earth is purely a onetime accident. That would mean, of course, that we are actually alone. While evidence points away from that conclusion, it does remain a possibility.

In Hawking's scenario, as life evolves there is an eternal struggle for survival with the best predators at the top of the food chain. Therefore, successful alien life-forms would be both fast and deadly. If true, that would be a concern for deep space astronauts arriving on foreign planets. Protective measures would certainly be advisable as such predators would definitely have the home field advantage. However, it is unlikely that spacefaring aliens would be so restricted to predation. It would be reasonable to apply *Maslow's Hierarchy of Needs* when considering the problem. If the ET was on the low end of social evolution and still concerned with survival issues, internecine conflict would be most probable. Therefore it would be unlikely that a subset of such a civilization would have acquired the skills necessary to develop deep space travel. In addition, human experience has already demonstrated that, concurrent to physics and engineering, advances in biological manipulation would allow the self-generation of food supplies. Scouting the sparsely populated reaches of the universe in search of sustenance is therefore a highly unlikely scenario. Note that we have both developed embryonic space flight, yet still engage in wars, and Hawking did mention the extermination hypothesis. That is, once a civilization develops nuclear capabilities, they eventually self-destruct. Hopefully this thesis is not correct.

Where Hawking really went astray was in his description of alien abductions and their relationship to UFOs. He began this narrative by stating, "The story is always the same—a lone individual on a quiet road at night, and taking an unscheduled detour." The reality is these cases did not begin in the 1950s, as he stated, nor is there just one scenario such as he described that encompasses the range of experiences. This is a rather blatant case in which a scientist failed to differentiate between his personal opinion and facts.

As has been noted, there have been descriptions of human-alien encounters as long as humans have recorded history. While a few episodes such as Hawking depicted have been reported, the preponderance of abduction experiences are initiated from a person's bed. The lack of perceived safety in what should be a secure location is one of the more terrifying aspects of these encounters.

While wildly off the mark regarding the facts of the abduction scenario, Hawking did raise a point on which I could agree. He appropriately implored viewers to consider these stories "from the alien's point of view." Why, he wondered, would an advanced alien race visit Earth and engage in seemingly random experimentation with test specimens?

The positive outcome of Hawking's documentary was to bring renewed attention to the concept of alien life-forms. His examination of their physical embodiment focused on a purely biological evolution based on an adaptation to their unique environment. Predation was the key issue. Not mentioned were the factors that have allowed humans to progress to our current state, including development of more complex brain functions or the simple but significant issue of an opposable thumb. However, as with my disagreement with Richard Feynman's conclusions regarding UFOs, it is a mistake to constrain the discussion and not take into account the evident aspects that consciousness plays a significant role in whatever we are observing.

Skinwalker Ranch

Extensive research in other equally arcane fields contributed to development of the notion of possible precognitive sentient phenomena. The organization called NIDS has been mentioned as was its mission to explore UFO sightings. Some readers also will be aware of a most unusual ranch in southeastern Utah that was acquired by Robert Bigelow. Sometimes known as *Skinwalker Ranch*, it has been a focal point for a wide range of phenomena for at least decades, possibly centuries. For a more lengthy discussion of the activities there, the book *Hunt for the Skinwalker* by Colm Kelleher and George Knapp is recommended.

It was the UFO reports that attracted Bob's attention to that property. However, what happened there did not fit into any neatly organized box. For consistency in reporting, I will use the same pseudonyms that Colm and George used in their book.

Orbs or balls of light were often reported by the Gormans, who worked the ranch. They were very troublesome and contributed to the decision to sell the ranch. In one memorable instance, three of the Gormans' dogs were observed snapping at the orbs as they floated through the eastern part of the ranch close to where they lived. These orbs seemed to tease the dogs and led them off into a nearby pasture. The dogs never returned to the house. When Gorman searched for the dogs the following day, all he found were three grease spots on the ground, which he took to be the remains of the dogs. Fearing that his teenaged sons might provoke the orbs, he decided to vacate the premises, selling it to Bob.

I was the first of the NIDS scientists to pull a watch at Skinwalker Ranch. Nothing unusual happened that night that I'm aware of or on any other occasion while I was present. However, in the years that followed many extraordinary events would happen with our highly skilled scientists observing. If there was one common factor to those events, it was that they had nothing in common with one another. As bizarre as many of the incidents were, they tended to be totally unique.

Worth noting was that in addition to our scientists who spent many cold nights scanning the area, Bob had the area instrumented as well. There were cameras strategically placed on the ranch taking time-lapse photography 24/7. Each camera snapped a frame every second and a third, day and night. These tapes were then reviewed and searched for anomalies. While unusual events were experienced on the ranch, none ever showed up on camera.

One of the most perplexing examples included a situation in which the camera on one pole was interrupted, yet the camera on a second pole that was oriented directly toward the damaged camera detected nothing. The damaged camera (camera one) was located on a pole about twenty feet above the ground. The camera observing the event (camera two) was on a similar pole. As all cameras were recorded with date and time stamps, the exact time of the damage was known. The extent of the damage was significant. The wire attached to camera one was ripped out and dangling loosely. There was about a three-foot-long segment of the wire leading down the pole from camera one that had been cut and was missing. Tests on the remaining segments of wire suggested that a rusty instrument had made the cuts. The wires leading to camera one were affixed to the pole by a large amount of duct tape (probably about half a roll). That duct tape was totally missing. Anyone who has worked with such tape can attest to how hard it is to remove that material. Further, near the ground, the wires had been protected from animals by being encased in PVC tubing and held to the pole via U-clamps. The PVC had been pulled loose from the pole and the U-clamps were again missing.

The videotaped observation from camera two was revealing in what was not seen. At the time of the event that is based on the stamp when camera one stopped recording, camera two did not display anything out of the ordinary. Coincidentally, the cattle just happened to have been grazing right around the camera one pole at that same moment. They did not move in any excessive or excited manner. That is important as we knew the cattle would scatter any time a person or predatory animal approached them. That ruled out that someone was able to sneak up on the pole by hiding behind the very narrow sector (a matter of inches) that was blocked from the camera two view. Considering the amount of physical damage that occurred, for the entire event to have happened in a little over a second (or between video frames) is simply out of the question.

Another albeit different example that presented an inexplicable set of circumstances involved the mutilation of a calf. This case is unique in that it happened on a sunny midmorning, in an open field with the rancher not far away, and within a very limited period of time. It was calving season and Gorman went out to check the herd. He found a newborn calf and as was customary weighed it and tagged the ear to identify the calf with its mother. He then drove about 300 meters across the flatlands of the ranch where he found a second newborn and repeated his procedure. The registering of the second calf took about forty-five minutes. Importantly, he was always within line of sight of the location of the first calf.

When Gorman drove back across the field, the mother of the first calf was going berserk. On the ground was the mutilated dead body of her calf. The ear containing the tag appeared to be surgically sliced off and was missing. The calf was eviscerated and completely exsanguinated, not to mention that about twenty pounds of the animal were totally missing. The body was generally intact, held together with one femur lying a short distance away. The bone contained markings that were later determined to be indicative of a sharp knife.

While UFOs have been periodically associated with these events, most cattle mutilations occur when no one is present and the exact timing cannot be determined. This case was different in many aspects. Every conceivable alternative scenario was considered, and none make sense if limited by normal physical reality. Predators, including bears, wolves, and large cats, were considered and ruled out. None behave like this. The lack of blood was problematic. As a test, blood was obtained from a slaughterhouse and poured onto the ground. Even weeks after the experiment it was very clear where blood had been spilled, yet no blood had been found in or near the calf. The notion that some team of people raced across an open field and was able to conduct this extensive amount of surgery in a short time is highly improbable. In fact, such an attempt to rustle a cow in that area of the country could be suicidal. Yes, they probably would shoot them on the spot. As inconceivable as it seems, the best explanation was that the mutilation actually took place at another location and the body returned. Only two alternatives allow for that scenario; an invisible UFO or interdimensional interaction, and both exceed our understanding. This is a case in which PSP proved to be completely in charge, and absolutely malicious.

A Creature from Another Dimension Enters the Ranch

One other case that Colm and George covered describes yet another set of highly unusual circumstances that can only lead to extradimensional possibilities. During August 1997 two trained researchers established an observation point on an escarpment that runs along the north side of the ranch. At about 2:30 A.M. they spotted a dim light that appeared in the vicinity of the road below them. Over time the circle of yellow light grew slightly in size and intensity until it was about four feet in diameter. Strangely, the light seemed to hover about three feet above the ground and a third dimension was noted. Rather than a simple circle, the light gave the appearance of a tunnel. Using our third-generation night vision goggles the researchers were able to see the events that unfolded quite clearly. A moving dark shape appeared in the tunnel and came to the surface. Then the humanoidlike creature of considerable size—an estimated six foot tall and four hundred pounds—using arms pulled itself out of the tunnel of light and landed on the road. Shortly thereafter the creature walked off into the darkness and the tunnel of light receded and disappeared.

Allowing a prudent amount of time, the two NIDS observers made their way down the cliff to the road below. As often is associated with Bigfoot or Sasquatch sightings, they noted a pungent odor in the area but found no trace of the creature. In daylight they returned to the site and spent hours looking for tracks that should have been inevitable with a beast of the size and weight described. None were ever found.

Some, but not all, interactions with strange animals took place at night where tracking was difficult. Prior to Bigelow's acquisition of the ranch, Gorman had an encounter that defies all logic. It took place at about noon and at extremely close range. What was at first believed to be a very large dog was spotted approaching from the area dominated by Russian olive trees on the west. As the animal got closer Gorman and other family members noted it now appeared to be a wolf, but one bearing the demeanor of domestication.

The size was most impressive as the head was chest high on Gorman who stands well over six feet. The wolf allowed petting and did not appear to be threatening to anyone.

As Gorman resumed work in the area he was quickly alerted to a problem by the bellowing of cattle in an adjacent pen. The wolf had reached under the bottom railing, grabbed a 600-pound calf by the snout, and was attempting to pull it toward him. Gorman grabbed a heavy wooden pole and struck the beast with several hardy blows. That was to no avail and the wolf continued its attack on the calf. Gorman ran to his truck that was parked a short distance away and retrieved his 44-magnum revolver. Then, at point-blank range he fired six shots into the area that should have contained the heart of the animal. That at least caused the beast to let go of its grip on the calf, but did not kill it as would be expected.

As the wolf turned and began to slowly trot away Gorman escalated the confrontation. His next weapon was a rifle he carried for hunting elk, which is an animal that can weigh six hundred to a thousand pounds. Gorman fired, hitting the beast with such force that it visibly tore chunks of flesh from the body. Apparently unconcerned, the animal continued to amble away from sight and disappeared. When Gorman found the residue that had been ripped from the body, he noted that it had the odor of putrefaction, something that does not usually occur for days after death. The next day they did observe a UFO in that area.

The Relationships Between Various Phenomena Are Noteworthy

Like cattle mutilations, the relationships between UFOs and creatures of unknown origin are frequently made. If there is to be a comprehensive understanding of these phenomena then it is imperative that we be more inclusive than exclusive of the data.

Of the cases already mentioned in this book, there are a few that appear to support the PSP hypothesis. Cash-Landrum is one of those worth considering. As previously indicated there was a physical craft that was emitting radiation. That led to injuries to the three people that exited their vehicle as the UFO approached. They all agree that there were a number of helicopters that were observed circling the disabled craft. More specifically, some of those helicopters were identified as U.S. Army CH-47 Chinooks.

Therein lays the problem for researchers that only allow for a resolution in our physical domain. While we know with great assurance that the people were irradiated, we have almost no evidence that the helicopters were real, and it is extremely unlikely that they did belong to the U.S. Army. On one hand researchers who are bound to consensus reality (that is, the physical reality that is experienced and agreed upon by most humans) come to the conclusion that if helicopters were seen and reported by multiple sources, therefore the U.S. Government must be lying and covering up responsibility for an accident—an argument that has been rejected in Federal court. On the other hand, if PSP is considered, then many more alternatives become available. The two that leap to mind include an ability to control what is observed. One possibility would be that the UFO could employ holographic technology to create the UFOs. Another alternative is that the UFO was able to project that imagery directly into the brains of the observers, thus actually manipulating their perception of reality. It is interesting to note that the phenomenon would allow substantial physical injuries to the observers. Based on this case, and a number of others in which physical injuries have occurred, the PSP must not be mistaken as benign.

The Bentwaters case also displayed characteristics that behaved like PSP. Less defined and rarely discussed are the other phenomena associated with this sighting. Of course an object under intelligent control that responded to the advances of the Air Force personnel is quite significant. Also reported on several occasions were strange noises coming from the woods. Farm animals in the area were also making quite a commotion. At other times there were also unconfirmed reports of ghostlike apparitions in the area. If true, this case has many of the same features as the events at Skinwalker Ranch.

The Interactions with Strategic Weapons Have Profound Implications

The difference in the cases involving U.S. and Soviet missile intrusion also offers insights in possible PSP. A major difference between the two countries' approach to command and control are noteworthy. The United States had a decentralized system. Once authority was given to launch nuclear weapons, the two launch control officers had the means to initiate the attack. The Soviets, conversely, did not trust their launch control officers to the same degree that the Americans did. Therefore, they had a centralized command and control system whereby launches were handled by Moscow. Launch control officers could not unilaterally fire their missiles.

Now consider what happened at each location. At both sites UFOs were observed. Obviously, a 900-foot craft hovering over Byelokoroviche, Russia, for several hours was more dramatic than the ones being reported by USAF security forces. In the Malmstrom case the missiles were shut down, yet at the Soviet site the missiles went through the early phases of preparing to launch. The similarity is that the officers in charge, and their senior leadership, lost control of their systems. Of course the good news is that in neither case were any injuries incurred.

The fundamental question now becomes whether a PSP was able to interact with those nuclear weapons in the manner most likely to cause concern. The implications of each interaction were significant in and of themselves. If, however, an agency (in these cases a PSP) was cognizant of the circumstances surrounding the operating parameters of those strategic systems, and actively intervened, then the implications are truly profound.

Step Back Proposal

For several years I have made proposals at conferences for studying PSP. This is called the Step Back process. Many scientists have agreed with the concept but to date there have been no takers who were prepared to provide the financing necessary. In fact, it would not cost a great deal. The abbreviated version is as follows:

— A list of participants is carefully developed. They should have broad expertise in their respective fields and be able to articulate exactly what people have sighted.
— The group is assembled at a sequestered location for a minimum of three days. Living together for those few days is an absolute requirement as the cross-fertilization of ideas is critical to the outcome.
— Initially participants are asked to list the observations that are made by people who have seen unusual events. No value is attached to the observation (for example, this was a UFO, poltergeist, ghost, Sasquatch, and so on).
— These observations are displayed on monstergrams that are posted where everyone can see them. Even with today's computer technology, these would probably be extremely large rolls of paper.
— Once the observations are all listed, the group is asked to step back where they can see all of the input.
— The next step is a macropattern analysis seeking to determine previously unrevealed relationships among the varieties of phenomena listed.

The prediction is that scientists involved will come away with a new understanding of the complexity of the problems we are trying to solve. With a little luck, there should be some reasonable designs on how to proceed.

Disclosure

There are many people who are demanding that the U.S. Government disclose the secrets it may or may not hold regarding extraterrestrials. The question to those making such demands is if other than for personal or prurient interest, then for what purpose? After all, why is it that these groups need external verification of something they believe they already know? In many ways belief in ETs is akin to religion. We do not ask governments to validate the existence of God or a Supreme Being. In fact, as a secular nation, the United States expressly separates functions of church and state. While polls indicate the country and the Western World are becoming increasingly secular, we ascribe religious beliefs as a matter of faith. Yet in this one area—UFOs—demands are made to validate individual beliefs.

If the issue is simply a *"gotcha,"* then the topic joins a litany of issues about which the government has been less than truthful. As such, it would be relatively trivial compared to unwarranted wars, a near collapse of the economy, and calamitous environmental issues, all of which have immediate life-and-death consequences.

More important, official disclosure has happened. However, for both skeptics and true believers, no amount of verification will be sufficient. There is a litany of senior officials who have already made disclosure proclamations, and to no avail. In America President Truman and President Reagan stated UFOs were real. Mikhail Gorbachev, President of the Soviet Union, stated they were real. Many senior military officers and high-level public officials from several countries have also claimed UFOs are real.

Given the lengthy list of world-class dignitaries who have officially announced that UFOs are real, what else could be done to establish confirmation of the facts? This is a far more complex problem than it might seem on the surface, and is not isolated to this topic. At the heart of the matter is how belief systems are formed and just how difficult they are to change, even when one is presented with overwhelming evidence. That is especially true if an individual has taken a public position on the topic. Admitting fault and correcting personal errors are among the most difficult things a person can do.

In reality, the disclosure movement is both a distraction and disingenuous. The proponents of disclosure have a predetermined agenda, and if documents fail to support their position, then the government, or *THEY*, are hiding the good stuff. It is a Catch-22. Besides, several countries have made disclosures and the results were less than earthshaking. The French Government put their files on the Internet. The system crashed from overload, but that just indicates a high level of interest. French society did not change. The UK government followed suit. Aware of the French experience, they conducted a timed release that worked better. While there were a few spectacular cases in those files, such as Torres, the vast majority of the reports were quite mundane. The British lifestyle did not change. Announcements by senior officials in Brazil, Chile, Mexico, and Russia that UFOs were real had no impact on daily life there or elsewhere.

Strangely, the people advocating disclosure point to those foreign releases as examples of transparency. While the foreign documents are suggestive of craft with unusual capabilities, none of the governments have claimed that they belong to ETs. However, those same advocates will accept nothing less than confirmation of ET contact by the U.S. Government. Somehow, it is believed that U.S. officials have special knowledge that no other government has. The CIA, U.S. Air Force, and the National Archives all have made documents available online. The near-universal response has been, "Not enough!" What proponents of disclosure really mean is, *"Confirm what I already believe—or else you are covering up!"*

For the record, I do not believe that there are pervasive secrets held by government officials that would change how people regard this issue. However, one recurring theme is worth addressing: that official acknowledgment of ETs would have a catastrophic impact on societies. Barring an all-out invasion in which ETs are coming to eat us, or otherwise subjugate all of humanity à la the *ABC* network television program *V*, the response is likely to be surprisingly muted. The public reaction would be based on two factors—the sense of personal threat and ambiguity. Evidence of demonstrated physical violence, or threat thereof, would evoke the strongest reaction. Ambiguity also would play a role in determining the response. In general, the more facts that are available, the lower the ambiguity, and thus the lower the perceived threat would be.

Obviously there is an extremely wide range of possibilities for announcement of acknowledgment of the existence of extraterrestrial life. These could include discovery of unicellular life on one end to face-to-face contact with sentient beings on the other. The response would vary accordingly.

On the lower end, the finding of simple life-forms or intercepted distant communications would probably rate a few news items for a day or two, both in print and television. Unless it was a very slow news day, they would not even lead. Those would be followed in academic circles with a number of scientific papers and debates at conferences.

Acknowledgment that craft of nonhuman origin exist, appear to be under intelligent control, and transit our atmosphere from time to time has already happened. Except in highly localized instances, these result in more brief news reports. Then those stories are responded to by skeptics who claim it is impossible and that there are prosaic explanations. The public follows these stories based primarily on their existing beliefs about UFOs. Basically this is the status quo.

Assuming there were no overtly hostile acts, if there were a highly publicized UFO landing, possibly with real-time television coverage, or a similar incident with confirmed interaction between ETs and humans, interest would increase significantly. However, the worst-case scenario regarding impact on society likely would be similar to what happened in the wake of 9/11. In fact, that would be a reasonable model to explore. That means a flurry of activity for several days. It would begin with nonstop television coverage, which would devolve within a few hours into breaks for commercials, and eventually routine programming would occur as other issues compete for attention. For several days it would impact productivity as people would stay glued to their television sets. However, within less than a week daily life would return to near normal.

Certain Factors Predicate Human Response — Material and Psychological

In all situations, the driving factors would be food and fuel. These are expendable commodities that are quite sensitive to fluctuations anywhere in the distribution system. In general, grocery stores and filling stations only have a few days' worth of inventory on hand. They are dependent on a continuous resupply through an intricate network that demands constant attention. As has happened during various crises, barring serious physical damage, the system would return to near-customary levels rather quickly.

How quickly normality returns would be primarily a function of the continuity of communications and the availability of water and power, both electricity and natural gas. These factors differ from food and fuel in that they are delivered directly to the consumer through established networks that require attention by a limited number of people. Obviously the presence of critical services such as law enforcement and fire-rescue are important, and unlikely to waver. The financial sector may see a blip and the stock market would experience short-term reverberation. The financial impact would be minor compared to the near-catastrophic failures recently experienced.

The potential impact that the confirmation of extraterrestrials would have on religion has been raised in several articles. There are those who believe that ETs would be incompatible with Christianity and feel a revelation would threaten the foundations of the church. Monsignor Corrado Balducci of the Vatican has written about the topic and does not see a cause for concern, while alarmists fear violence might erupt. Rather, Monsignor Balducci stated, "It is very well possible that other inhabited planets exist. We do not find any direct reference to extraterrestrial life in the Bible, but it neither excludes their existence. Since God's wisdom and omnipotence has no limits and is infinite indeed, this possibility doubtlessly exists."

There have been surveys conducted in which clergy were asked directly about how they and their parishioners would respond to the news of an advanced ET civilization. One of those was a project run by my wife, Victoria Alexander, for the Bigelow Foundation. One thousand Catholic, Protestant, and Jewish clergy were contacted and the response rate was very high. Contrary to dire predictions, they indicated it would not precipitate any crises. Victoria noted in her conclusions, "As the *Alexander UFO Religious Crisis Survey* illustrates, religion should not be summarily categorized as unresponsive or inflexible to challenging matters. They do not appear to be in danger of disintegration at the news that the UFO community believes would overwhelm them." She goes on to state "that religious leaders did not believe their faith and the faith of their congregation would be challenged by contact with an advanced extraterrestrial civilization—one with or without a religion. According to many respondents, it would confirm God's glory as the creator of the universe."

There is yet another hypothesis worth considering and that is, *"What if they are us?"* For many years people have speculated about the possibility that humans were seeded here on earth. This concept suggests that somewhere in the evolutionary process, ETs stepped in and through genetic manipulation created what became the human race. This allows for both evolution and intervention. While highly unlikely, if it were revealed that testing had demonstrated alien engineering of DNA, and that had been previously known and concealed, the public in general and scientists in particular would be very upset. While disclosure of that information would be troubling, it would not lead to societal disorder—just a lot of soul-searching and tough explaining. Of course, that opens up another line of thought. If ETs did alter the evolutionary system, "For what purpose?"

The bottom line on disclosure is that it is not the single most important issue in the world that the UFO community makes it out to be. In fact, it pales compared to many existing earthly problems. The question posed by Victoria that puts the issue in perspective is, *"Does this mean I don't have to go to work or school today?"* Yes, you do!

Roswell

Of all of the UFO cases, Roswell remains the most mystifying as the pieces never quite fit together. Of course, having the U.S. Air Force balk so badly was a great disservice to the American people. Even the General Accounting Office report notes that something unusual happened near Roswell in July 1947. Attempting to come to any complete resolution of the incident is a fool's errand as whatever the truth may have been it has become so distorted over time that facts have merged inextricably with fiction. In the end we are left with little physical evidence but considerable alleged eyewitness accounts. Interestingly, while eyewitness testimony is known to be highly unreliable, in the case of Roswell it suddenly seems sacrosanct. A bad choice.

Roswell is still an outlier, that is it is an event that does not fit anyplace on the normal scale of events. There are several alternatives that seem to make more sense than a crashed UFO. First of all, the location must be suspect. The alleged crash site is located in an area just outside of White Sands Missile Range (WSMR) that incorporates Trinity, the site of the first atomic bomb test. The Roswell Army Air Field was immediately adjacent to the eastern boundary of that rather huge area. In 1947 *Project Paperclip* was in full swing and many of the participants located at either Redstone Arsenal near Huntsville, Alabama, or at WSMR. Paperclip, of course, was the use of German rocket scientists who would eventually put us on the Moon. It should be noted that the early U.S. Army rocket programs did not go well. At WSMR there were many attempts at launches that ended up never even leaving the pad, or were intentionally destroyed shortly after liftoff. It was so bad that unsuccessful Army rocket launches became the butt of many jokes.

Considering that America would be going in to space a decade later, it is worth considering what those scientists would have been doing in 1947. In research and development vernacular, they would have been working on enabling technologies. That means development of basic technologies that later will be incorporated into the more advanced systems. As weight was a vital factor in all space-related activities, they would have been researching stronger, lighter, and flexible materials, much like those reported by people who saw the wreckage. I note that the GAO report ruled out my first guess, which was an errant rocket launch.

The incident was also discussed with Colonel Bob Friend, U.S. Air Force (retired) and former head of Project Blue Book. When asked about what happened at Roswell, his guess was a *Broken Arrow*. Most people will recognize that as the code name for a nuclear weapon that has been lost without risk of going to war. There have been more such incidents than is commonly known to the public. An accidentally dropped weapon was a consideration, given that at that time Roswell Army Air Field had the only nuclear-equipped bomb wing in the entire world. The GAO report also rules out that scenario. Further, in checking the limited nuclear inventory at the time, no devices were reported missing. Either of these situations would have caused the kind of response that was indicated by people associated with this incident. They certainly would explain some pretty heavy-handed actions that seem to have been taken by the military.

Material Was Recovered

Everyone agrees that something was recovered in Mac Brazel's field. What that was remains in question. The notion of the Project Mogul balloon appears to have more credibility than I would have thought just a few years ago. The testimony of Major Jesse Marcel has always been impressive and most researchers believe that he was telling the truth as he knew it. Years later I had the privilege of meeting his son, Dr. Jesse Marcel, Jr., who has had an exemplary career both in the military, where he became a helicopter pilot flying in Desert Storm, and as a civilian medical doctor. His modesty is noticeable as is his ability to stick strictly to facts. As he states, his UFO experience took place during a matter of a few minutes more than sixty years ago. He never embellishes on the information, as many other UFO researchers are apt to do.

In a recent interview with a knowledgeable source, the discussion turned to the composition of the elements of the Mogul balloon experiments. Among the materials involved were thin reflective metal pieces cut into an exact configuration to reflect a specific frequency. What was learned was that on the reflecting panels had been placed a specially designed code that could only be read by the people with access to the key. More important, it was stated that this code was not alphanumeric as are most that are frequently employed, but entailed the use of glyphs. That was very important information as Jesse Marcel, Jr., frequently has described seeing such markings on the material his father brought to the house.

There was an experiment conducted by Bill Birnes on his *UFO Hunters* television program that is also worth noting. In it there were two witnesses who claimed to have seen the crash material, Jesse Marcel, Jr., and Staff Sergeant Earl Fulford, U.S. Army (retired). What was done was to place several sheets of thin metal material on a table. Then each of the witnesses was independently asked to come in and see if they could identify the material that they saw from the Roswell crash. Strikingly, both Marcel and Fulford chose an acetate substance that had most of the qualities reported by Jesse Marcel, Sr., in his early reports. It was lightweight, strong, and flexible. When crumpled and released, it would return to its original shape and will not burn. In his analysis of this experiment, Birnes dismissed the possibility that acetate was used in aircraft at the time. Frankly, I am loathe to rule out a known earthly material that was readily available in favor of an extraterrestrial crash without a lot more hard evidence. But more important, the acetate that was identified on the program has the characteristics required of the panels. Mogul balloons went to very high altitudes and frequently were buffeted by heavy winds. That type of material was necessary so that the panels could return to their original shape after wind deformation.

Another credible firsthand witness that has come forward is Jack Townsend, who at the time of this event was a lieutenant working for Major Jesse Marcel, Sr. In a recorded interview he repeated and confirmed much of the same information provided by Jesse Marcel, Jr. There is no reason to doubt his observations, but it can be accounted for by Project Mogul material. Given the extreme sensitivity of that program, none of the witnesses had been given access to the information. Additionally, the Air Force knew that Soviet scientists would see the press release from Dallas, thus switching material was a logical step in protecting the program.

Project Mogul makes sense for several reasons. While the Air Force report, *Case Closed*, provides conflicting information regarding classification, most of those involved agree it was both Top Secret and strictly compartmented. This fits with the concerns of the times and how almost everything related to nuclear weapons was treated. Based on the little-known concept of the high-altitude sound channel, Mogul was designed to fly extremely high, near the stratosphere, and allow us to listen for a nuclear detonation coming from the Soviet Union. It is true that they did not achieve success until a few years later, but we only knew they were working on the bomb, and not how close they were to successful detonation at that time.

"It was a different time." Given the extremely sensitive nature of this project, the response that has been reported by witnesses is commensurate for the time. By today's standards cajoling or even threatening witnesses seems untoward, but those who remember "duck and cover" will understand the situation and be more forgiving. It was indeed a different time and at the beginning of the Cold War rather ugly. The notion to do whatever was necessary had

real meaning at that time. The possible inadvertent revelation of such a classified program can account for many of the alleged anomalies, including the sudden movements of senior officers.

Security oaths have often been used as an excuse for not telling the truth about the Roswell incident and it does appear that some people were verbally threatened at that time. However, that is no longer the case and anyone having knowledge about the Roswell incident has been released. Prior to the interviews by the U.S. Air Force, Secretary Widnall released those persons from any previous security obligations that may have restricted their statements.

Considering Walter Haut's Statement

One of the more troubling questions has been to account for the affidavits that were left by Walter Haut. As a lieutenant, Haut had served as the Public Information Officer at Roswell Army Air Field who had released the statements concerning the UFO crash recovery to the press. There seems little doubt that Walter Haut was an honorable man. Interviewed many times, Haut in 1993 issued a sworn statement concerning the events of July 1947. Then, in 2002 he wrote another formal sworn affidavit that was released only after his death. There are, however, significant problems that arise when attempting to reconcile the two accounts, as the later version places Haut as a firsthand witness, something not previously claimed during his live interviews. The 2002 affidavit confirms a solid egg-shaped craft and the existence of alien bodies.

The following is Walter Haut's 1993 affidavit in its entirety. As will be shown, this can be reconciled without ET and shown to be consistent with the facts available.

AFFIDAVIT

1. My name is Walter Haut.

2. My address is: [————BLACKED OUT————]

3. I am retired.

4. In July 1947, I was stationed at the Roswell Army Air Base, serving as the base public information officer. At approximately 9:30 A.M. on July 8, I received a call from Col. William Blanchard, the base commander, who said he had in his possession a flying saucer or parts thereof. He said it came from a ranch northwest of Roswell, and that the base intelligence officer, Major Jesse Marcel, was going to fly the material to Fort Worth.

5. Col. Blanchard told me to write a news release about the operation and to deliver it to both newspapers and the two radio stations in

Roswell. He felt that he wanted the local media to have the first opportunity to have the story. I went first to KGFL, then to KSWS, then to the *Daily Record* and finally to the *Morning Dispatch*.

6. The next day, I read in the newspaper that General Roger Ramey in Fort Worth has said the object was a weather balloon.

7. I believe Col. Blanchard saw the material, because he sounded positive about what the material was. There is no chance that he would have mistaken it for a weather balloon. Neither is there any chance that Major Marcel would have been mistaken.

8. In 1980, Jesse Marcel told me that the material photographed in Gen. Ramey's office was not the material he had recovered.

9. I am convinced that the material recovered was some type of craft from outer space.

10. I have not been paid nor given anything of value to make this statement, and it is the truth to the best of my recollection.

/s/ Walter G. Haut

Signature witnessed by: 5-14-93 Max Littell. /s/ (Date)

All of the statements made by Haut could be true, even without a crashed UFO. Note that every statement concerning the UFO was based solely on information provided to him by other officers. Remember, at the time of this incident he was a lieutenant. This was not a period in which a very junior officer would question the word of someone as senior as a major, let alone a colonel—and there was no need to. Given the tight compartmentalization of information about Project Mogul, it is unlikely that either Colonel Blanchard or Major Marcel would have been briefed into the program. They would not have had a need to know; a critical component to be included in briefings on the topic. That General Ramey had the material switched and the weather balloon story concocted also makes perfect sense as it is a typical cover story to block further disclosure. As this was a high-profile breach of security, they knew the Soviets would be watching closely for our explanation. Thus the only comment that we are left with is Haut's opinion that he believed the object was a spacecraft. He was entitled to his opinion, but as he stated, it was based on the observations of others. Unfortunately that does not alter reality.

Like many of the readers I have heard so many conflicting stories that it is hard to tell what to believe. None of the information fits with the reports of recovered aliens, dead or alive. Interestingly, Dr. Jessie Marcel, Jr., provides considerable credible information regarding the material he observed, but makes no mention of alien bodies. It does not seem plausible that his father would show him the debris yet fail to mention the most important aspect of the incident. Conversely, lunar astronaut Edgar Mitchell has told me that he does believe alien bodies were recovered. As previously indicated, such information appears to be contradicted by the concerns of NASA regarding encountering microbes on the Moon.

I still remember a discussion I had with Bruce Maccabee regarding the debris. One of his poignant comments was in asking why they had not found even a screw that was identifiable. Of most concern to me is that there is scarce evidence that the alleged recovered material went to the places that legend has them. In the end, it may be that Roswell is another example of a PSP. Explaining Roswell is never easy, and given the high degree of cross-contamination of information, the truth will probably remain forever clouded.

Alien Abductions

One of the phenomena closely associated with UFOs is the question of abductions of humans by aliens. These were often described as nocturnal affairs in which the victims were taken against their will to another location where they were subjected to examinations by an assortment of beings. The accounts by already established author Whitley Strieber are probably best known. His bestseller and landmark book *Communion* catapulted the abstraction of these events from idle curiosity of small groups into mainstream public consciousness. The cover depicting a gray alien with a bulbous head and soul-sucking, deep, dark eyes became an icon for the movement. There were a host of other books including *The Andreasson Affair, Interrupted Journey*, and the books by artist-turned-hypnotist Budd Hopkins, who assisted victims with recovered memories. Dr. David Jacobs, an abduction researcher at Temple University, focused on the negative aspects of the phenomenon as he addressed the incidents in terms of a threat against humanity.

The entry into the arena by Dr. John Mack, a Pulitzer Prize–winning author and renowned Harvard University psychiatrist, elevated the discussion considerably. However, his personal experience at Harvard demonstrate the animosity that research in phenomenology can engender. Despite being a tenured professor, after his first book in the field, *Abduction,* was published, the Dean of Harvard Medical School appointed a committee of peers to review Mack's clinical care of patients though he was not suspected of ethics violations or professional misconduct. In fact, it was the first time in Harvard's history that a tenured professor was subjected to such an outrageous investigation. While he was eventually cleared, the investigation consumed fourteen months and brought about considerable unnecessary stress.

I first met John at a conference on the phenomenon that he and Dr. Dave Pritchard held on the campus of Massachusetts Institute of Technology (MIT) where Dave is a prize-winning optical physicist who developed the field of atom interferometry. Like Harvard, MIT was concerned about their image when they realized what the conference topic was. Everyone in attendance signed a statement that this was an abduction conference held *at* MIT, but not *sponsored by* MIT. As the agenda worked out, I spoke immediately following John's transformative presentation and addressed the commonalities between near-death experiences and alien abduction. He was a tough act to follow.

Contact with both John and Dave was maintained for the following years, albeit for different interests. Once at NIDS more involvement took place as some assistance was provided to John by Bob Bigelow. Over the years I had several private conversations with John. While his public image was that he was naïvely accepting all of the fanciful tales of the abductees at face value, he was in reality much more circumspect about the topic. We both agreed that there were factors at play that we did not understand and that the abduction reports had underlying aspects that defied simple explanations such as the bad aliens were coming and snatching people without consent. Again, John Mack was not naïve.

Such interactions between humans and nonhuman entities have been recorded from time immemorial, and certainly long before UFOs became popular. Like UFOs, there are tantalizing indications that some tangible evidence points to the physical reality of at least some of these events. That all the people who claim to have been abducted by aliens are crazy does not fit the evidence. State-of-the-art research using fMRIs suggests that identifiable markers in the brain can be located that confirm there is something physical to these events. This evidence is commensurate with the observations I have made of shamans all over the world. However, exploring that is another book.

Puzzles Addressed

Alien Bodies

Hangar 18 at Wright-Patterson Air Force Base is legendary for housing the alien bodies that were acquired at Roswell. While there were no cadavers stashed there, there was a UFO sequestered in another hangar at the base. Again it was Colonel Bob Friend who informed us about the facts behind these stories and another knowledgeable source confirmed the data. There was a UFO, but it was built by humans and located in another building. There was a need for secrecy as well.

The UFO in question was actually a training device that was used in educating young intelligence officers. As part of the course work, a night session was scheduled. With no other preparation, the trainees were taken to the darkened hangar. Upon entering the lights would suddenly come on and the students would have a brief period to observe the craft. They were then led to another location whereupon they were debriefed about what they had observed. The point of the training was to prepare them to gather details when confronted by a totally unanticipated circumstance. For most people, when surprised they focus on very little and quickly forget the essential elements of information. This exercise was designed so that if these officers were ever in a situation where they unexpectedly had access to an advanced enemy's system they would not be overwhelmed but instead mentally take notes as quickly and efficiently as possible.

Obviously surprise was a very important factor. Over time there were a few people who stumbled on to the site by accident. At that point they were debriefed sans many details and forced to sign security statements related to inadvertent disclosure of the material. It would appear that over the years a few people may have avoided being caught. The rumors about UFOs held there are quite consistent with such a set of events. According to the sources, the model was moved to another location and its current disposition is unknown.

Some of the rumors may have been exacerbated by the onetime director of the Foreign Technology Division at Wright-Patterson, Colonel George Weinbrenner. Weinbrenner was a World War II pilot who had been shot down over Belgium, and with the help of resistance fighters, he managed to evade capture for seven months. A MIT graduate, he later spent a considerable amount of his career involved in technical and scientific intelligence. His friends knew that he was wont to joke about holding "little green men in the cooler." It is apparent that some people were not aware of his jocular proclivities.

The Government and Cattle Mutilations

There has been considerable speculation by some conspiracy theorists that cattle mutilations are actually performed by agents of the U.S. Government. One reason for such a suggestion is the technical sophistication involved in some cases. For instance, in a case examined by George Onet, a NIDS scientist with double doctorate degrees, including a Doctorate of Veterinary Medicine, a cow was found downed in a most peculiar manner. While there was a small hole in the frontal neck area, the pericardium was intact, and yet the heart was pulverized. The extent of injury was commensurate with a high-powered rifle, yet there were no concomitant wounds of entry and exit. Of course, if a bullet had been involved, the heart would not have been still encased. The best guess was that some form of a high-powered and directed energy was used. The rural setting made that improbable, at least by humans.

According to the rationale provided, the government's purpose for the killing and mutilation of these animals is some form of surreptitious surveillance program. This of course fails the test of logic. Should any government agency wish to monitor the environment via animal studies, it would only have to buy them. If anonymity was desired, then a third-party contractor could obtain the necessary specimens. This approach would be far more cost-effective and risk-free. The method proposed by the conspiracy crowd is fraught with danger, both personally and institutionally.

The collusion conspiracies don't stop with cattle mutilations. According to the most extreme conspiracy theories, the nefarious U.S. Government has made agreements with evil aliens and traded advances in technology for unwitting human victims to serve as sample experimental material. Some conspiracy theorists state that the aliens trade technology for human DNA. Preposterous— there are many less intrusive methods to obtain DNA and they would have solved cloning long before now.

Secret Aircraft—the Flying Triangles

In many parts of the world sightings have been made of UFOs that appear as large flying triangular-shaped objects. NIDS investigated more than 700 reports, both in the United States and abroad. There has been speculation that these are really secret aircraft developed by the U.S. Government. In fact, during the many sightings that took place over Belgium in 1989 and 1990, they asked our forces with NATO if they did belong to us. The short answer is no.

This is another situation in which common sense can be brought to bear. While sometimes these triangular UFOs are silent and darkened, on other occasions they appear brightly illuminated. Some of the sightings may be attributed to the V-shaped B-2 Spirit, but certainly not all. As an example there is a famous picture of a triangular craft over Belgium that clearly shows a lighted craft in the night skies.

One of the best documented cases in the United States took place over Illinois on January 5, 2000. At about 4:00 A.M., Melvern Noll, a small business owner in Highland, Illinois, spotted a large, brightly lit UFO moving slowly to the southwest. He notified the local police department and officers confirmed the sighting. By all accounts the UFO was huge. One observer stated it was a massive elongated triangle the size of a football field and about two to three stories high. Others had it larger than a Boeing 747 aircraft or a C-5A. The height from the ground was estimated at about 200 to 900 feet and it flew relatively close to some of the people making reports.

Due to the direction of movement, the Highland Police Departments notified other law enforcement agencies in the area. Successively the object was seen by officers from the towns of Lebanon, Shiloh, and Millstadt, Illinois. All observers confirmed the slowly moving gigantic craft that had bright lights on it. Some observers stated the object was silent, while others heard a faint humming sound. After the UFO passed Millstadt, it made a large U-turn, headed northeast, then with a sudden burst of speed went silently zipping off into the breaking dawn.

Exhibiting high-strangeness/high-credibility the case was sufficiently well documented that two television specials were done about the incident. More witnesses came forward and it seems other police officers saw the craft as well. The investigation indicated that the Air Force was not particularly helpful, other than to state it was not an aircraft that belonged to it.

This case made enough news to catch the attention of the *Los Angeles Times*. The article described the research efforts by various organizations. The writer, Stephanie Simon, included one side comment that speaks volumes about how the phenomena are perceived by local people. She stated, "Folks out here don't seem too taken with the mystery. They tend to be pragmatic sorts: they worry about their jobs and their farms and the weather, not giant flying arrowheads. Some think it was an alien craft. Others nod knowingly: secret military project. Either way, it doesn't much affect them."

During the investigation I had several e-mail contacts with archskeptic Phil Klass. He went public with his resolution to the mystery—Venus! Ignoring the facts yet again, his response to a large bright object observed over a period of time at distances as close as a few hundred feet was that he wanted us to believe they all had misidentified a tiny distant planet.

There are several significant problems with the notion that these large aircraft were produced by the United States. The first obvious factor is that no one would fly a secret plane over a densely populated area with it totally illuminated. That is antithetical. The transition from speeds that would stall conventional aircraft, followed by rapid acceleration that exceeds known capabilities, suggests technologies that are unknown to modern aviation. The dates of sightings are also problematic. Some of the large triangles were routinely observed in the early 1980s and would certainly be known by now.

I have had the opportunity to discuss heavy lift with a friend of mine in the

Air Force. A major general (now retired) at one time, he was in charge of the Air Force research and development at Wright-Patterson AFB. At another point in his career he commanded Luke AFB, the location of the famous Phoenix Lights case. Logisticians know that heavy lift capacity is the long pole in the tent for many military operations. When participating in what was known as the Chief's Wargame, at Maxwell AFB years ago, he also learned just how limiting our current capabilities are. That is despite all of the C-141, C-5A, and C-17 aircraft in the U.S. Air Force Air Mobility Command. The reality in Iraq is that our forces could not function without the materiel supplied via civilian air transportation companies.

Major General John is aware of my interest in UFOs and we have discussed the topic on several occasions. He is quite skeptical, but at least open to talking about the subject. Aware of the allegations about large triangular craft, I asked him directly if they could be ours. His definitive answer was that we have nothing larger than the aforementioned aircraft. That answer makes sense from a pragmatic standpoint as well. If we had larger transport aircraft, they would be widely known throughout the military. Vehicles and equipment are not just thrown on airplanes willy-nilly. There are intricate loading plans. When vehicles arrive to be airlifted for long hauls, they are carefully marked for center of gravity, height, and weight. Each vehicle has a loading plan and every item inside is placed according to that plan. That means that thousands of lower-ranking enlisted people, many without any clearances at all, would be briefed on the process.

While there is conclusive evidence that large triangular aircraft do transit our airspace, you can be equally sure they are not ours.

Summary

This chapter addresses many complex yet important issues. Whatever the origin of the UFO phenomena, it is unlikely that a single simple theory will adequately resolve all of the observations. Such involution appropriately has led to differences of opinions among scientists researching UFOs, and none has a corner on infallibility.

The self-anointed skeptics willfully choose to ignore veridical evidence as it does not fit their preconceived concepts about what is possible. It is important to remember that anything that does happen, can happen. That some facts are inconvenient or exceed theoretical limitations is irrelevant.

Clearly, UFO sightings are but a subset of a multifarious macrocosm of phenomena. These, when examined in a broader context, suggest that emergent patterns are discernible. These seem to intimate that there exists sentience that lies beyond current comprehension, yet controls the presentation of seemingly improbable observations.

Importantly, those who advocate public disclosure of official data pertaining to UFOs need to examine their own motives. It appears that like the skeptics, they have a preconceived notion of the information and what they really desire is external confirmation. The need for validation is primal to the human psyche. In reality, official disclosure already has been made by several countries, and there is no reason to believe that the American Government would add anything substantially different to that body of knowledge.

While there are many documented cases that have both high strangeness and high credibility, there are many other stories that emerge from a grossly distorted mental morass. Too frequently these tales gain popularity, and the stranger the better. Some of them are considered next.

THE TWILIGHT ZONE

There are two things that are infinite, the universe and human stupidity;
and I'm not sure about the former.
—ALBERT EINSTEIN

When it comes to fantasies and conspiracies regarding UFOs, there appears to be no outer limits that bound reality. At many UFO conferences those who present the most bizarre stories are best received. Sex with aliens is a best seller, especially if the details are as titillating as they are outlandish. Nocturnal encounters with reptilians is a favorite subject with audiences.

Delving into the psychological components of these reports would take at least another book. But it is worth noting that tales of human-creature sexual interactions easily predate the advent of UFOs. What is striking about many of the presentations is the strong emotional component attendant to those making the claims.

In 2008 while attending Victoria Jack's annual UFO conference in San Jose, California, I was accosted by a rather irate older woman. Although she thought herself to be a victim of an alien abduction, she also blamed me. According to her, she had witnessed me on board a UFO supervising experiments on those captives there. While patently untrue, this incident personalized just how caught up people can become in believing in situations that could not possibly happen in consensus reality—that is, the physical world as we know it.

We have already addressed the notion that the U.S. Government or the related *THEY* have reverse engineered a UFO. We have also noted that even distinguished figures, such as the former Minister of Defense from Canada, believe that consensual interactions between extraterrestrials and human governments have taken place and that they openly comment on the topic. There also exist bizarre narratives involving ET interactions at underground bases at Area 51 and other locations. Some of these unbounded stories are worth exploring, if only so the uninvolved general public can know about them and have enough information from which to draw their own opinion. One that hits close to me is covered next.

MILABS

There are a whole host of reports concerning what are called military abductions or MILABs. Unfortunately, as the lady mentioned earlier indicated, I have personally been associated with many of the stories. There is some crossover drawn here between my interest in UFOs and work with nonlethal weapons. In fact, on more than one occasion I have been picketed at conferences. Some of the people voicing objections see nonlethal weapons leading to mind-control systems. They are joined by other self-proclaimed victims who form the connection between earthly mind control and alien abductions. In fact, there is not a great deal of separation between many of these groups.

Interestingly, one of the proponents of these abductions was allowed to make a presentation at the *5th European Symposium on Non-Lethal Weapons,* in Ettlingen, Germany, on May 13, 2009, and I was given time to respond. While no violence was projected, the local *polizie* were present to ensure protests remained peaceful. In preparation for rebuttal a startling statistic was uncovered. Between 0.5 and 2.5 percent of the population can be characterized as schizophrenic. In numbers that means that in the United States alone a minimum of 1,500,000 people, most of whom are undiagnosed, fit that category. Stated that way, the number of truly bizarre incidents that people report being engaged in is more understandable.

Mind control has a long history. The current mind-control Web sites often address the latest breakthroughs in monitoring capabilities. However, timing is a problem for them. The first such device, the *Air Loom,* was designed in 1810, long before electromagnetic influence became popular. Interestingly, the concept came from a patient in Bedlam, a world-renowned mental institution. Of course technology has advanced and the microchips that are currently available could allow for new capabilities. However, the alleged MILAB victims claim this monitoring has gone on for more than fifty years.

To address the MILAB topic, the following article was written by me in 1999, but has never been published formally. The concepts are as valid today as they were then and are offered for careful consideration. This was in response to the April 1999 MUFON article, *"Military Interrogation Sessions with Alleged Abductees,"* by Helmut Lammer, Ph.D.

In response to the various comments, I am submitting this commentary on Military Abductions (MILABs). As with the first response, and contrary to assertions by Helmut Lammer, this commentary addresses only his MILABs, not all abduction cases. Interestingly, most of the respondents who have been whining on the Internet do not even address the question at hand. That is human, not alien, interventions.

It should be noted that Victoria Alexander was the sole author of the article entitled *"What Would Freud Say?"* while I provided information about satellite systems. Contrary to the comments about the timing of the article, it was written in 1996, within a month of Lammer's first piece in the *MUFON Journal*. At that time it was sent to the editor, Dennis Stacy. In a phone conversation with him, Stacy stated there were more important issues to cover. He was subsequently relieved as the editor of that journal. From the April issue it is clear that MUFON has little interest in discussing both sides of this controversial non-UFO issue. Instead, it has chosen to become the champion of unsubstantiated, barely tangentially related nonsense.

While rejected by MUFON, Victoria's article has been circulated privately for the past three years. After a recent meeting in Laughlin, Nevada, and upon his request, Victoria sent the article to Peter Gersten, who posted it on CAUS. This actually provided a wider audience from that of the moribund MUFON, which has been in steady decline for several years. It is only through coincidence that the article came out in proximity to the nonsensical book being published by Lammer.

Again, Lammer has displayed both a lack of understanding of technical knowledge and an inability to competently analyze information. Unfortunately, he is not alone. This topic is quite important and is covered in some detail in my forthcoming book, *Future War, Non-Lethal Weapons in Twenty-First-Century Warfare* (St. Martin's Press, May 1999). However, given his current position in the Austrian Space Research Institute, this obvious lapse is very disconcerting.

While triangulation technologies have existed for some considerable time, the devices had a common characteristic. They were relatively large. Until very recently, they were certainly larger than could be surreptitiously subcutaneously hidden in a person's body. They would be too large to go undetected by X-ray or MRI examinations. By 1990, the state of the art was a device about eleven millimeters long and two millimeters in diameter. This was a passive transponder without an internal power source. While it could be read by external devices, they had to be very close, similar to readers that are currently used in stores. Greater distances can be achieved with higher power or a very narrow bandwidth. That means the device would have to be larger to house internal power or once externally interrogated the information rate very low. It would also have to broadcast above the background noise at the designated frequency. Today, most of the miniaturized location technology is designed to work at a range of feet, not miles. The trade-offs between, power, frequency, antenna systems, and size make the commonly accepted notion of MILABs highly implausible. Since it is claimed that these military abductions have been taking place for quite a period of time, it must be assumed that older technology was in use. It is noted that Lammer's response to Victoria's article posted by CAUS does not reference a source earlier than 1996. Yet he wants us to believe these nearly nanoscale, mystical capabilities have been available for decades. They have not.

Attempts to locate persons who are free to move about at substantial distances infer that the interception capability is quite mobile. Therefore, it would be logical to assume space-based systems, or airborne platforms, are involved. As previously noted, the space-based systems were not available when the incidents began. However, his article does mention that "MILAB victims are harassed by dark, unmarked helicopters" that are seen in the area. It seems incongruent that abductions must take place discreetly, yet helicopters are sent to openly harass abductees.

Even if this mystical and unattained technology were available, the organizational aspects are illogical. If the three-car system proposed by Lammer were used by the offenders, the logic still fails. Assuming there are a minimum of two people per car, and we know that it takes five shifts to man any given position, we are led to assume that thirty people are assigned to continuously track each MILAB. Of course, that doesn't count supervisors and administrative personnel. Remember the helicopters. Where did they come from? Who flew them? Who conducted the medical tests? Who maintained these yet-to-be-identified bases? The list of involved personnel goes on and on. Since it is claimed that these illegal operations have been conducted for many years, and since military personnel rotate on a frequent basis, there would have to be thousands, if not tens of thousands, of people involved over time. Where are they? When you add up all of the people who say they are MILAB victims, the number of people involved in this operation would be nontrivial. At a time when military strength is cut to the bone, are we to believe this mission-without-purpose takes precedence over other critical functions? With all of the questionable projects that have been exposed, why have we not heard from one whistle-blower about MILAB? The reason is because it does not exist, nor has it ever existed.

It must also be noted that not one implanted transmitter has been recovered. According to the MILAB theory proposed, location and extraction of such a device would be a simple matter. None of the "alien implant" work, such as that done by Roger Leir, has found anything remotely associated with a human designed transmitter. The systems described in Lammer's article for use in monitoring criminals are quite large when compared with some unknown subcutaneous version and are generally tracked by fixed sources. This is hardly what the "abductees" have described. Where are these devices?

Had such technology been available (and it was not), then the analogy between abductees and Gaddafi, Noriega, and Saddam Hussein would be apropos. None of these people rose to prominence out of nowhere. They were all identified on their ascendancy, and at a time when there was physical access to them. If available, triangulation transponders could have been implanted long before these foreign leaders became problematic.

How current is the analogy of MILABs versus critical national interests? On March 31, 1999, three American soldiers were captured near the Yugoslavian border in Macedonia. The incident is making news around the world. It is noted that their exact location at this time cannot be determined. Are we to assume that some über-secret agency believed it was more important to use this highly advanced technology on unsuspecting civilians—who are of no special interest except to themselves and their friends—than to use it in support of our national security? That logic is certainly seriously flawed.

Lammer's knowledge of "black programs" and how they function is equally lacking. His argument appears to be that if large amounts of money were available, some agency, or even a renegade subelement thereof, would choose to spend money on tracking innocent people. Here there are two key issues to address. First of all, just because a project is "black" doesn't mean there is no oversight. While fewer people have access to the program, in this day and age, dollars are watched quite closely in every project. His notion of the money available is off by at least an order of magnitude. Lammer states that SDI was funded at "tens of billions of US dollars." The reality is that at its zenith, SDI was a five billion dollar program and that has shrunk dramatically over the years. When I proposed a politically sensitive project to Lt. Gen. Jim Abrahamson, then director of SDI, he turned me down. He stated that if he were caught funding that venture, Congress would assume SDI had too much money and would make severe cuts. In fact, SDI was cut by one billion dollars that year. What I had proposed was nowhere near as risky as illegally kidnapping civilians and physically assaulting them.

The second major problem in his thinking is that such a project makes no sense to anyone—except for a few conspiracy theorists. There is no logical purpose in tracking the people who make these claims. They do not appear to have any significant attributes that would make them worthy of special study.

There is no indication that they are extremely intelligent, nor is their physical prowess of note. No Nobel laureate or person of publicly acclaimed accomplishments has ever claimed to be a MILAB. It therefore remains a mystery as to why these relatively nondescript people are reportedly chosen to be unwilling MILAB participants.

There are, however, many well-known medical conditions that describe these signs and symptoms. These observed or perceived contentions, which maintain that some person, or group of persons, is after them, are indeed found among ten million persons per year who are seen by clinicians who include social workers, clinical psychologists, neurologists, general practitioners and other primary care practitioners, and sometime by psychiatrists. The vast majority of these persons (more than 90 percent) function very well within the activities of daily life. Outside of a narrow, highly circumscribed paranoid delusion, if only a concern or worry, even if not real, these delusions are not disabling. Most common are beliefs of other entities, voices, bedroom-related visions, and hypnagogic experiences. Often they are accompanied over time by exaggerated notions of self-importance. Clinical estimates are that more than 100 million of these persons are alive today worldwide. They are found in all countries and all cultures. Generally speaking, the demography of these persons does not match that of a normal population. These observations are not new. They have been diagnosed, or more likely merely observed, for more than a century. It is a much more simplistic answer than the bizarre scenario being portrayed by Lammer and his supporters. Where is Occam when you need him?

The reason that no agency would engage in such a preposterous program is the potential for repercussions versus value added. We can find no value added by unauthorized, illegal monitoring of individuals who are self-proclaimed MILABs. After all, a volunteer program would net better results and at no risk. In this day and age, any agency caught conducting such outrageous experiments as has been postulated would risk both severe personal and organizational consequences. My educated guess is that the organization would be disbanded and individuals sent to jail. Again, a simple cost-benefit analysis completely destroys the logic of conducting such an illegal project.

Both Lammer and Wilson draw analogies between the MILAB victims and unwitting participants in prior unwise government experiments. It cannot be denied that the system has been abused in the past and that individuals' rights were violated. However, in each of the cases listed, a reason for the experiments could be made. In Tuskegee, doctors wanted to determine how syphilis would progress if left untreated. Most, but not all, participants in MKULTRA were volunteers who signed statements to that effect. Forgotten in the clamor over those experiments is the grave concern that had been generated by our POWs, who showed signs of "brainwashing," when they returned from North Korean camps. The radiation experiments also were conducted in a time of extreme anxiety about the effects of exposure and were based on the concerns for our very national survival. I am not making excuses or apologies for these experiments. However, in each case, the designers conducted a risk-benefits analysis and chose to proceed. The proposed MILAB projects fail that simple test of common sense. There is just no reason to conduct them. However, since there is no statute of limitations on kidnaping, I also highly recommend that any person who experiences abduction at the hands of anyone, including purported government agents, report them to multiple law enforcement agencies. That will ensure that no single agency can quash the report. While immediacy would be preferred, old cases also can be filed.

Surprisingly, Victoria has been attacked for both taking information out of context, and using quotes that are quite long. The intent of quotes was to show the words were accurate. The reason the quotes were extensive was to ensure they placed the situation in context. Again, this is an example of internally inconsistent logic by the critics.

Finally, the MILAB concoction fails every known test of knowledge, proof, and common sense. Not one scintilla of concrete evidence exists to support the hypothesis. Lammer's arguments fail in technology, political science, military science, government, budget and finance, organizational sociology, and psychology or psychiatry. No one supporting the MILAB hypothesis can explain why critical resources, if they existed, would be employed for this nonsense versus some issue of vital national importance. At the end of the day, all we are left with is abstruse, totally unsubstantiated conspiracy theory. But that does sell well in some circles.

In the intervening period and a decade after his article and book appeared, Lammer, an Austrian astrobiologist, has published a retraction stating he no longer believes his earlier material. Unfortunately he did not state what he now believes. However, the notion of MILABs has not gone away and the connection to abductions by extraterrestrials has been reinforced. In 2009 James Bartley wrote *"A Primer on Military Abductions,"* which specifically addresses this issue. In it he states, "The reptilians are the primary abducting force. The reptilians use 'proxies' such as certain factions of Grey aliens, Nordic, Insectoids and other ETs to do their bidding. The reptilians are also deeply involved in Milab Operations." He also brought up claims made by an alleged National Security Agency employee, Dan Sherman, who has made outrageous assertions about his efforts on that agency's behalf. Sherman noted that in his NSA assignment he was "trained as an 'Intuitive Communicator,'" who was responsible for establishing and maintaining communication with extraterrestrials. He also indicated that there were "interactions between deep black elements of the Military-Aerospace-Medical community with non-human beings."

The claims get more bizarre from there and we learn that some of the MILABs and their perpetrators are shape-shifters, and others are being treated by military psychologists who are aware of the situation. Bartley notes the "most violent milabs are reptilian or drac [a winged gargoyle species] human hybrids." Of course his next step is time travel operations, which he indicates are already ongoing.

Bartley is not alone. In May 2010 George Knapp forwarded me an e-mail that said in part, "I must also, Caution [*sic*] you that Col. John Alexander is one of those Reptilian Shapeshifters who eats humans for breakfast, works for the Devil and is an impostor not the same person people have come to love and trust."

Dulce

Snuggled near the northern border of New Mexico is the quaint town of Dulce. The population is less than 3,000 and mostly comprised of Jicarilla Apaches. To the immediate north rise steep mountain escarpments and remote areas that few people visit. For UFO buffs there is a unique combination of real sightings, strange phenomena, and abject nonsense. Like the Uintah Basin, Utah, home of Skinwalker Ranch, the area around Dulce is also steeped in historical mysteries.

Among the real events that have occurred in the area were a substantial number of yet to be explained cattle mutilations. One of the main investigators of those cases was Gabe Valdez who was assigned there when he was with the New Mexico State Police. The events were very perplexing and Gabe went out of his way to assist the local ranchers in attempting to determine the source of the brutal attacks. Gabe later assisted NIDS in our investigations of activities in the area.

Investigation of the mutilation cases proved to be frustrating for Gabe. He sent many samples to the Colorado Bureau of Investigation (CBI). The response was always the same—predation. In order to test the veracity and thoroughness of the CBI, Gabe ran his own experiment. He and a rancher killed a cow intentionally and cut it up using their knives. They then sent samples to the CBI for analysis. The response, of course, was predators. It was at that point he knew he could not trust the "scientific evidence."

As a member of the New Mexico State Police who sometimes patrolled the highways late at night, Gabe had several UFO sightings of his own. He described one occasion in which he saw a glowing UFO that was more than a hundred feet across open a side panel and emit a smaller dark craft. The second object then flew off to the east. In another case involving observation at relatively close range, he states he saw a UFO flying right toward Mount Archuleta. He expected to see a crash, but instead the UFO just disappeared. It seemed to fly directly into the mountain and did not reappear.

One of the myths that can be put to bed involves a report of a crashed UFO that was recovered by a U.S. military unit. Gabe told us he had been on patrol along Route 64 one night when he encountered Chinook helicopters fully loaded with troops. As the only law enforcement officer in the area, he attempted to ascertain what they were doing. The commander, a full colonel, was decidedly uncooperative and basically told him to mind his own business. The helicopters then took off and disappeared into the Archuleta Mountains. Rumors that a UFO had crashed were rampant, which was just fine for the military. Those rumors continued for years. In reality a prototype stealthy aircraft had crashed and the recovery team was there to pick up every scrap of material that could be found. No one, not even law enforcement, was let in on the real operation.

Another example of high-strangeness/high-credibility events was related in a previously mentioned book by Kelleher and Knapp. It was quite dramatic and originally told to us when Colm had come back from one of his trips to the area. The witnesses include nine people traveling in four vehicles. Two of them were senior officials from the Jicarilla Apache tribe. The event took place in 1996 at about 11:30 P.M. and under excellent conditions for visibility. As they were driving through a canyon, the occupants of the cars saw a huge object that was moving slowly toward them. It eventually went directly over them at an estimated 100 to 200 feet in elevation. The size was dramatic. It was described as so large that as it flew it overlapped the edges of the mesas rising on both sides. That would make it more than a mile in diameter. Without making a sound, the object suddenly moved off at great speed. There are many other reports on unusual incidents that both defy explanation yet seem to have a basis in consensus reality.

That is not true for all of the stories that emanate from that area, and they get personal. According to mythology there is an underground base hidden deep in the mountains. This base was used by both humans and aliens, but the people who lived in the area could never figure out how they entered the facility. Here is an example of the material written about this secret base:

> Most of the aliens supposedly are on levels 5, 6 and 7 with alien housing on level 5. The only sign in English was over the tube shuttle station hallway which read "to Los Alamos." Connections go from Dulce to the Page, Arizona, facility, then onto an underground base below Area 51 in Nevada. Tube shuttles go to and from Dulce to facilities below Taos, N.M.; Datil, N.M.; Colorado

Springs, Colorado; Creed, Colorado; Sandia then on to Carlsbad, New Mexico. There appears to be a vast network of tube shuttle connections under the U.S. which extends into a global system of tunnels and sub-cities.

What is most amazing is the total disregard for even basic facts. The comments about the intercity underground connections that are ascribed in these myths are worth noting. It would appear that the authors have no concept of just how hard tunneling really is. An interesting comparison is to examine the Chunnel, the connection between Coquilles, Pas de Calais, France, and Folkestone, Kent, United Kingdom. This spot was chosen as it is the shortest distance between the two countries. In World War II a great ruse was accomplished to make the Germans believe that the real invasion force would come across at this location. The distance is less than thirty-two miles and it took six years to construct at a cost of more than twenty billion dollars. Even allowing that the engineers had to bore under the English Channel, this task pales when compared to the distances of many hundreds of miles alluded to in the Dulce fantasy.

One can also explore this in more detail by looking up *The Dulce Papers*. They describe the base activities and note that the aliens live on blood and are here to alter our DNA. Allegedly there are other underground bases involved in the experimentation. As you might guess, one is allegedly located near Groom Lake in Area 51. As for the true purpose of these efforts, the paper says, *"Only God, MJ-12, and the aliens know for sure."*

The Taos Hum Is Real and Unexplained

However, some of these tales play out in reality, and without explanation. For several years a phenomenon called the Taos Hum made headlines. It seemed that a small number of people living in the area could hear a low-pitched humming sound, even though 90 percent could not hear it. With both Sandia and the Los Alamos National Laboratories located within about a hundred miles, scientists did check out the reports. Much to everyone's surprise, their instruments could detect a hum that is below the hearing threshold of most people. The problem was that the source could not be determined.

One day I received a call from a woman who claimed that she was bothered by this sound. She also indicated it had plagued her to the point that she moved from the area to another state. The problem was that the sound did not abate over distance. While we know acoustic waves can be transmitted over long distances, they usually diminish as you get farther from the source. As was pointed out to the caller, that alone should tell her that it wasn't the national labs that were creating the problem. Other more conspiracy-oriented people believed these sounds were coming from an underground base or from the imaginary tunnels mentioned previously.

At War with the Aliens

The ridiculousness does not stop with simple stories of secret subterranean bases. Here is an e-mail, one of several such, that actually accuses me of direct involvement in fighting the aliens.

> To: UFOFacts@yahoogroups.com
> Subject: Re: [UFOFacts] DULCE base!!!?
> Date: Mon. Aug 25, 2003, 20:06:45 EDT
> The Dulce Base was actual, but the subject of enormous and distortionary disinformation. The real (govt.) base was beneath Archuleta Mesa. It was a location for working on back-engineered Star Visitor technology. The disinformation would have you believe that evil alien invaders did dastardly experiments at the Dulce Lab. Such is not the case. The Black Hats, including Army Intelligence Colonel John Alexander, were involved in nuking the Lab.

What is scary is that the author of this e-mail holds a real doctorate degree and was a licensed therapist in the State of California. According to him, the secret government *"Cabal operatives use high technology, including antigravity 'tractor beams' to draw persons up into disguised, silent, hovering antigravity helicopters masked as 'UFOs,' which then take the victim to an installation made to seem like a UFO base, so that the victim will become confused and blame 'the aliens' for Cabal kidnapping and abuse."* In addition we have other technology as "the Cabal can render the victim unconscious by either psychotronics or sleeper gas."

He is not alone in discussing wars that have taken place between humans and aliens. As indicated in the e-mail, there are quite a few people who believe that a major battle took place near Dulce and that the U.S. military used a nuclear weapon to destroy the aliens.

That author has posted a list of The Good, The Bad, and The Ugly regarding UFOs. As seen here, I lead one group:

BAD GUYS/WOMEN (Bad = harms UFO truth—a partial list)
Colonel John Alexander, USA, DIA, INSCOM, (ret.), Ph.D.,

head of Bigelow's NIDS; member, Aviary; high-ranking officer in the UFO Cover-Up Cabal

A sense of grandiosity often accompanies these tales. Consider this e-mail that was widely broadcast in which the author elevates his position to speak for all of humanity:

> Greetings. Permit me to introduce myself. I am Dr. Richard Boylan, appointed by Star Nations as Councilor of/for Earth with respect to Star Nations High Council. I will be speaking now for both Star Nations, and for the Human species of Earth.

Obvious common sense plays no role in these fantasies. Why would we reserve these advanced technologies (that we don't have) for the purpose of abducting people of no particular skills or interest, yet not use them in known combat situations? Of course, there is no reasonable answer. The purpose for including this outlandish material is to allow readers to understand just how bizarre the UFO business can get. These are a sampling of why one needs a sense of humor if they choose to participate.

But these examples are not as benign as they might first appear. The entire field of UFO research, including the best studies that have been conducted, is clouded at best. Determining fact from fiction is always difficult. The problem with these bizarre representations is that they provide fodder for the skeptics/debunkers that use the stories as examples of what the entire UFO community believes. Remembering the discussion about elected officials and their reluctance to comment on UFOs—it is precisely because they know future opponents will link a rational position to these outrageous ones. One thing the twilight zone does prove is that smart and crazy are *not* mutually exclusive characteristics.

Summary

The fantasies associated with UFO phenomena are boundless. There is nothing that can be contrived that will not find a willing audience. Unfortunately, many people who have woven the tales have created an exotic tapestry, but one that embodies the properties of flypaper, or possibly a Venus flytrap.

Therein lays an important downside for serious researchers. These outrageous tales provide fodder for ad hominem attacks that employ fallacious logic. Despite the fatal flaws, guilt by association is established and reputations—no matter how eloquently established—can be demolished quite rapidly.

There are two key factors worth considering. First, the number of people who can be categorized as having serious mental health problems is far greater than generally imagined. The second, however, is more perplexing. That is that reports of contacts between humans and nonhuman entities have existed for time immemorial. The inclusion of UFOs as an integral element of these observations is relatively new. It would be easy to attribute all such phantasmagoria accounts to mental illness. However, such behavior would bear resemblance to that of the UFO skeptics. Ultimately, scientific research does require that problems be identified. Still, there exists the underlying issues of what constitutes the boundaries of reality.

Between the domains of high-strangeness/high-credibility and the twilight zone there is ample room for conjecture. The epilogue considers all of the information.

EPILOGUE

QUO VADIS?

Both the man of science and the man of action live always
at the edge of mystery, surrounded by it.
—J. ROBERT OPPENHEIMER

There is little doubt that some unidentified flying objects are real, three-dimensional solid objects, which are physically present and observable. To support their tangible attributes they are registered on a variety of sensor systems. They are seen by competent witnesses, photographed, have radar returns, are captured on FLIRs, leave radiation traces, and sometimes are spotted by surveillance satellites. If there is a problem in contemplating what UFOs might be, it is that wide mélange of observations and reports. With sightings ranging from small orbs of light to gigantic material craft miles in diameter, and every size and shape in between, it is almost impossible to define the phenomena or to establish bounding parameters.

Hard data confirm craft with extraordinary characteristics that defy current human technological capabilities. These include unparalleled acceleration and maneuverability—such that no living organism could withstand the g-forces—advanced stealthlike signal suppression, intradimensional manipulation of space, interaction with our strategic weapons systems—all under apparent intelligent control. Clearly there is no single theory that will explain the totality of such complex events.

However, *this is not the greatest story never told* as the true believers would have us imagine. Over the past six decades myths and conspiracies have enveloped the core elements and shrouded the narrative in fantasies from alien enslavement to the embodiment of our salvation. But where does that leave us today? The high strangeness of events, coupled with near-irrational objections from many in the scientific community, have rendered a technically competent study of UFOs close to impossible without great personal risk to one's career. The same is true for politicians who would like to obtain valid information. Given the current toxic environment, few elected officials would place future electability on the line just to satisfy the curiosity of a relatively small constituency who view UFOs as a high priority.

For good reasons, the UFO question is not a voting issue. The general population is interested in, but not committed to, the topic. To gain support for UFO studies at the Federal level, the subject must be able to compete favorably with such consequential matters as high unemployment, an unstable economy, record foreclosure rates, unpopular wars, unaffordable health care, nuclear proliferation, and global climate change. Seen in that light, it is obvious that UFOs will not become a front-burner issue—ever. In addition, the imaginary *THEY* don't exist, and we can forget about a POTUS waving a magic wand that confirms what the protagonists want to hear.

As has been elucidated, most of the information there is to disclose about UFOs is already in the public domain. Unfortunately, because the people have been lied to about so many important matters, the perceived integrity of the U.S. Government is so low that no amount of evidence or commentary from official sources would be viewed as credible. Foreign disclosure efforts, while interesting, have not had a significant impact. Certainly there has been no societal uproar. There is little reason to believe that further American disclosures would cause any additional repercussions.

Moving Forward

There are some actions that can be taken that will move us toward understanding the issues. Some require support from others, but the most productive ones will emerge from within the general public. If a change in perceptions about UFOs is going to come, it will be the spreading of the current grassroots efforts. Instead of looking to others for validation, it would be useful to have every person who has had a UFO sighting openly discuss it with others. The volume of reporting would likely increase the feeling that these events are widespread and more common than currently anticipated. In short, make UFO sightings a safe topic for discussion.

There is an urgent need to support scientists who are willing to investigate phenomena, including UFOs. That includes eschewing ad hominem attacks. In addition, while skeptics play a useful role, they must be held to the same standards they expect of researchers. They should not be allowed to make seemingly authoritative statements about events they did not observe without demonstrating that their solutions fit all of the facts, not just those of convenience. A skeptical position based solely on deductive reasoning is potentially fatally flawed. Such rationale can only be applied when *all* of the assumptions are supported by the evidence—not just those facts that seemingly fit the theory. In the study of phenomenology, inconvenient observations are too frequently disregarded. They should not be and thus being skeptical of the skeptics is essential.

This skepticism also should apply to the media who too frequently are willingly complicit in attacking the credibility of witnesses. Hoaxes will not stop, but they can be dealt with in the same way as any fraudulent claim or action that inhibits scientific investigations. If the acts are criminal—including filing false official reports—then treat them as such. Sending flaming balloons aloft is dangerous. Destroying the property of others is a crime. Any activity that adversely impacts safety needs a response. There is a need to let hoaxers know their acts are not as benign as they believe them to be. They should be held accountable. For example, when search and rescue or other public services are involved, send the hoaxers a bill.

The discontinuities between the government and the public, and the public and science, are so great as to be nearly irreconcilable. During the past few decades science and scientists have become more politicized than ever before. The fiscally motivated debates of global climate change and health care have proven that beyond a shadow of a doubt. In general the public believes that the scientists are all lying, and often what they say is based on who is paying them. Altruistic science died some time ago.

Governmental transparency is a significant problem that has not been altered despite changes of administration. The tendency of the military to classify everything because they can't figure out what should be protected continues to breed distrust. There is, however, a simple step that could be taken regarding UFOs. In conducting the last review of the Roswell incident, then Secretary of the Air Force Sheila Widnall issued a directive specifically releasing every witness from any prior security restriction regarding that case. A similar order from the Office of the Secretary of Defense would go a long way to clearing up the mess. Importantly, it would prevent many of the provocateurs in the field from hiding behind that smoke screen.

It is fully understood that the Department of Defense does not view UFOs as a threat, and therefore not in its bailiwick. However, that same department desperately needs public support for all of its endeavors. What the senior officials must understand is that a substantial majority of Americans believe that it is withholding information about the topic. Simply stating, *"We don't do UFOs,"* is totally insufficient to satisfy those accusations. Therefore, making a public statement that affirms that no information is being withheld, while releasing all witnesses from perceived restrictions, makes sense. First, it will cost virtually nothing. Second, it provides a great deal of goodwill. Third, it again affirms that the Defense Department is not the organization to turn to in reporting future cases.

It is recognized that the first part of the statement will not be believed by many people, especially conspiracy theorists. That is why the second piece, overt acknowledgment of freedom to discuss the topic, is the operant imperative. That privilege to talk deflates all of the arguments about aliens hidden at Area 51 and similar fanciful tales. For DoD executives it is time to set personal opinions about UFOs aside and engage in enhancing public trust. Obtaining goodwill at little cost should be a win-win situation.

While covered previously the issue of discussing UFO material in public is so important as to be worth reiterating. Anyone having information and who believes restrictions have been imposed on them should take note of the substantial number of government witnesses that have voluntarily come forward and made statements. My experience, and that of Robert Hastings, has been that no adverse legal action has ever been taken against a person making claims about UFO encounters. Again, the point is: *It is safe to talk.*

For those who wish to pursue obtaining additional governmental support in studying UFOs, I wish you well as you tilt at windmills. However, before launching your ventures you are strongly encouraged to learn more about how the U.S. Government really works. The efforts to date have been extremely naïve. Accusations and testimony at confrontational escapades may attract some media attention, but fall far short of a diplomatic call to action. Should you choose to have the Executive Branch authorize a formal UFO study, caveat emptor: The result could well be Condon II, rather than revelation of useful material. That is the most likely outcome and would set the investigation back several decades.

The same admonition applies to disclosure advocates. You must stop saying that *"knowing we are not alone in the universe will change everything." It won't!* Most people already believe that there probably is intelligent life elsewhere. Knowing that they still go to work every day and daily life changes little. Confirmation of what is already believed is not a paradigm shift, it is only interesting information.

For those who believe that UFOs will assist humanity in cleaning up our mess there is bad news: *ET did not answer the phone, the rescue ship did not arrive in time, and one is not likely on the way.* Therefore, we had all better get to bailing. The world is in bad shape and needs all of the helping hands it can get.

In order to better understand the complexity of the phenomena that are observed, again recommended is the Step Back approach outlined earlier. It is suspected that macropattern analysis will provide clues to the fundamental principles that underlie the observations. *This is basic science at its roots* and has the potential to develop a real paradigm shift—one that challenges a reductionist approach to all problems.

Where Does That Leave Us?

Based on credible witnesses and backed by physical evidence, I conclude that the UFO observations are manifestations of issues that are anfractuous and beyond current comprehension. The extraterrestrial hypothesis is but one possibility, and probably not the best fit with the facts. Unfortunately, it is the solution most frequently espoused by officials and scientists newly acquainted with the phenomena. If not from here, they must be from there is a logical initial conclusion. As noted, the wide variety of encounters suggests that postulate is inadequate.

We must beware of the *Puzzle Paradox* that plagues the field. That is the notion that we understand the issues and then collect data to affirm what we believe. In the past that has been done too frequently by enthusiasts and skeptics alike. With UFOs we are still collecting various disparate pieces, but there is no picture on the box cover that will assist us in solving this contumacious puzzle.

The UFO evidence offers a prime example of *Alexander's Law of Appropriate Complexity* that states: Every time one believes you fully comprehend your situation in life an entire new order of complexity is encountered. This holds true for every person and for every situation. Just as you have it nailed, something else comes along and changes the game. With UFOs, every time we think we have an answer, new observations make the problem more complex. Evaluation of the characteristics of UFOs and related phenomena suggests the extraterrestial hypothesis is too narrow, and once again the paradigm shift eludes us.

In the end it is clear that the universe is far more complex than we ever imagined. We are not close to solving the enigmas posed by UFOs, rather we are still on the front end of defining the fundamental issues and boundaries.

Letter to Lt. Col. Corso

9 September 1997

Lt. Col. Phil Corso
███████████ Drive
Port St. Lucie FL 34983

Dear Phil:

As I said to you when we talked last, as I went through your book I had a lot of questions about various facts reported. In order to be able to answer these questions as they are raised by others, I need your help and input. Some have been asked before by others, some are new, but both are included. Both questions of major importance and trivia are included. At this point, while extensive, the list is not all-inclusive. In general, I have put them in chronological order as they appear in the book. I have listed the page number on which the question was raised, but I have not included paragraph or lines.

> **Page 29.** Why would five trucks vs. one plane be used to move such valuable cargo from one air base to another?
>
> **Page 31.** It is not true that a duty officer would be authorized to go anywhere on base. Typically, many areas would have special security requirements the SDOs, or FODs would not hold. For example, Crypto, or SCI clearances would not be held by most officers, therefore they would be precluded from access to certain areas.
>
> **Page 32.** The opposable thumb has been key to primate advancement. Why would that not hold true for other species?
>
> **Page 33.** Why would shipment of bodies go to Wright-Pat. vs. straight to Walter Reed? It seems that with the potential for further decay, you would move them to the autopsy site as quickly as possible.

Page 44. Whose careers were pulverized? Can one be identified?

Page 55. In the 1950s we ate C Rations. MREs did not come along for a long time after that (1970s or 1980s).

Page 55. How could information be controlled if there was no central agency in charge?

Page 61. Of the two listed, the Skunk Works developed stealth proto-types (Have Blue) and the F-117. Northrop developed the B-2.

Page 63. Infers that FTD (Army) existed in since the 1940s. In fact, it came into existence when you arrived in 1961 and disappeared after you retired.

Page 66. States Paper Clip began to move German scientists to the U.S. a year before the end of the war.

Page 72. If this were "bigger than the Manhattan Project," where are all of the people who must have been involved? Over decades they would have numbered in the tens of thousands.

Page 72. It is stated that ET was killed by gunshot. Nothing that I know of in the literature supports that statement.

Page 72. If the Germans were reverse engineering from an ET crash, what happened to the scientific evidence? Would not that have come along with Paper Clip or been captured by the Soviets?

Page 77. You state that ad hoc groups throughout different branches of government participated. If so, how was information controlled? It would have leaked a long time ago.

Page 77. "the contact we maintained with the aliens," is a statement that infers we had (and still have) some form of communication with ET. Where is it? Also, is this not in conflict with comments about being at war with ET?

Page 78. It states that someone's job is to "manage our ongoing relationship with alien visitors." Again, there is no evidence that such an ongoing relationship exists.

Page 78. It states that reporters were given briefings with "truthful descriptions" and fell on the floor laughing. Who gave those briefings and to whom? There should be records available someplace.

Page 79. TYPO. I think you mean "our meetings were never FOR-MAL" in the second line.

Page 82. The number of security clearances should not have been a serious problem. Temporary read-ons are done routinely when necessary, so anyone needed for research could be included.

Page 84. The comment about Gen. Trudeau deciding to make international, if not interplanetary, policy in a vacuum should be very troubling. It certainly runs counter to the American notion of civil control of the military. I should think that senior officials from both the executive and legislative branches would be seriously disturbed by these comments.

Page 87. The CIA seems to have been grossly overstaffed if they had sufficient excess people to put tails on midlevel staff officers in the U.S. While a number of military people have retired and gone to the agency, I have never heard of them being harassed into joining. Rather, they went and applied. Given your vehement anti-CIA stance, why would they try to co-opt you?

Page 93. You state that two EBEs were alive after the crash. You had not mentioned that in any of our prior conversations. Can that be supported?

Page 99. Now both the Army and Navy have bodies to autopsy. Why would they be split up? It would make far more sense for one group to handle all of the work. That would allow for comparisons and minimize the security problems.

Page 100. You have alien craft going to Norton & Groom Lake. You also state that "the Army cared only for the weapons systems aboard the craft." That seems strange as it was the Army who was leading efforts into space and had the precursor to SDI. They therefore would have been interested in many technologies including propulsion.

Page 105. Here you state there were very strict rules about security clearance, yet you told us the security level was relatively low, but security was handled through an "old boy net" that only passed the info. to people they trusted.

Page 106. Von Neumann died in 1957. How could he have been involved in 1961?

Page 114. Is this ELM or ELF?

Page 114. We never had a "stealth fighter." The F-117 was actually a bomber.

Page 115. Is HARP actually HAARP or is this another program?

Page 115. If Von Braun went on record about UFO technology, where are those records now?

Page 117. How does a Lt. Col. in the Pentagon get his own staff car? When I was there, few generals had their own assigned staff cars.

Page 120. The crypto school would have been for NSA not the NSC.

Page 120. There has been an Air Force intelligence unit hidden in plain view at Ft. Belvoir for many years. However, their mission was not UFOs.

Page 121. Why would a society so technically advanced that we can't figure out how their systems work, worry about our defenses? When engaging forces with inferior systems we simply overwhelm them. I expect ET would do the same.

Page 121. Why would a sophisticated enemy worry about "large-scale warfare." With such advanced technology, one could bring an adversary to their knees without direct assault.

Page 122. Here the EBEs mean us harm. Yet on page 78 we had an established relationship with them.

Page 122. States that the Cold War was a cover for preparing to fight ETs. This implies a coordinated effort between the U.S. (plus NATO) and the USSR to maintain an earthly balanced threat so that defensive systems could be built to fight in space. The development and coordination of such an effort would have to have been massive. It would entail education of at least tens of thousands of officers to develop and execute the order of battle. In no way would you let a surprise attack catch an entire armed force off guard.

Page 123. Based on the previous statements I doubt "the Cold War was a cover for the secret agenda against ET." Can this be reasonably explained?

Page 124. How could even the most advanced fighters counter an ET threat? Compare the difference between U.S. and Soviet technology as seen in the Gulf War. The difference was probably a decade (our advantage over Soviet), yet we established air supremacy within two hours of initiating the attack.

Page 124. The same holds true for air defense technology. In hours U.S. and allied forces took down a very sophisticated air defense system that had been provided to Iraq by the former Soviet Union.

Page 129. No pictures taken on the Moon show an ET presence. Two of the twelve lunar astronauts are personal friends and confirm this statement. One is even supportive of the ET hypothesis but firmly states that nothing they saw on the Moon indicated an ET presence. Further, as the officer responsible for his, and late Apollo missions, he states there was never any indication of a potential ET presence or threat.

Page 129. SDI was NOT related to an ET threat. Again, several key people are personal friends or have other relationships with me. We have discussed the topic of UFOs. Even those favorably disposed to UFOs would state unequivocally that there is no relationship between SDI & ET.

Page 129. It states that surveillance techniques were developed to look both in and out (for potential threats). Actually, some satellite systems use Startacker to establish their relative position. This system lines up the stars, but does not survey outer space.

Page 129. Using overhead systems to find ET bases would be impractical unless they knew where to look. The coverage is just not good enough to find ET hiding in uninhabited areas. That would be especially true if they had advance camouflage or signal suppression capabilities.

Page 129. It's Adelphi, not Adelphia, Maryland.

Page 130. Why would a SP4 be surprised to see you out of uniform? Traveling in civilian clothes is done all the time.

Page 131. How did the Army get a sketch of a night vision device from the Roswell crash? Earlier you state the concept came from eye coverings taken from ET at the autopsy.

Page 132. Why would a UFO have seams? Certainly they would have been more structurally efficient than that.

Page 132. If the occupants of the UFO were basically robots, why would they need view ports in the hull? We already know that multi-spectral sensors would be more efficient in examining the exterior surroundings.

Page 132. Now the bodies are autopsied at the 509th, not at WRAIR (Walter Reed Army Institute for Research) or Bethesda as stated earlier in the book.

Page 136. You state that "Project Corona" was on your desk. Yet the recently released Corona documents state that reports were only made orally so there would be no paper trail. In fact, the only written document of Corona's authorization by Eisenhower was a note on an envelope by the DCI.

Page 136. Considering the stated animosity between you and CIA, why would they allow you near their crown jewels? I do note the ORD was one of the few Army offices aware of Corona.

Page 136. If overhead was being used to find ET, why was Art Lundahl, the director of the National Photographic Interpretation Center

(NPIC), left out? He was interested in UFOs, but never saw anything in all of the photos they collected.

Page 137. You state the mission of the U2 included looking for ET. I discussed ET/UFOs with Ben Rich, one of the Skunk Works presidents, on several occasions. He certainly did not know about this mission and he was interested in developing the most advanced surveillance systems.

Page 137. Again a problem with finding ET landing sites.

Page 139. Now we have sharing of Corona data with the Soviets? There is absolutely no way the U.S. would share our most advanced intelligence with them. In fact, the TK system was developed to protect the overhead photographic data. Another system still protects the satellite capabilities. These early warning intelligence assets were viewed as key to our national survival. Almost no one got direct data from them. When release was made, the photos were often altered to protect the actual capabilities.

Page 139. Corona was not build by LADC but another Lockheed company, LMSC, was the lead contractor. LADC built airplanes while several other companies actually built the Corona rockets and satellites.

Page 140. The resolution of the cameras aboard the U-2s and KH-4s operating in the 1960s was modest at best. The notion "because of our high-resolution aerial surveillance," would deny (see through) ET camouflage seems unlikely.

Page 142. The satellites coverage described does not fit reality. We still do not have worldwide coverage (1997), let alone what was available in the 1960s. In fact, one of the complaints about intelligence during Desert Storm (1991) was the times when we could not cover critical areas in combat. But, your passage has continuous coverage of "remotest parts of Asia, Africa, or South America" as early as the 1960s.

Page 142. I find the notion of "parity" with a society that is superior in technology hard to understand. Time and again, we have seen a technically superior force decimate a larger force that is less well equipped. In fact, our defensive strategy is based on the premise: "Fight outnumbered and win."

Page 142. Has "we" involved in Corona. This was a CIA and Air Force project. While the Army was a "user" I find no record of them being involved in Corona development.

Page 142. There is discussion of "reverse engineering Discoverer." While Discoverer was used as a cover for Corona, it was not a reverse-engineering project. Do you mean the payload was retrofitted?

Page 145. Comments here infer some form of treaty or alliance between the Soviets and ET. Is there any proof of this? We certainly did not see their military take any technical advances based on external assistance. In fact, the Soviet technology approach was almost always brute force. They did not even have the basic computer technology to keep pace with the U.S.

Page 149. The concept of placing a fighting outpost on the Moon does not make a lot of sense except as a possible trip wire. The outpost would be easily overwhelmed, and would be impossible to reinforce or assist should a battle take place there. Further, there would be little reason for an ET armada, traveling vast distances, to stop off at an intermediate point.

Page 156. What evidence is there of "EBEs who tried to scare us away from the Moon"? And, what is the evidence for ET bases there? As previously stated, lunar astronauts state there was no evidence of ET.

Page 157. States we sent enough manned missions to the Moon to challenge ETs. There were only seven lunar landing missions, including the ill-fated Apollo 13. We have not been back in twenty-five years (as of 1997). This is hardly a commanding presence.

Page 160. How would scientists at APG (Aberdeen Proving Grounds) marvel about a chip that came from the crash when they would not have had access to the crash or the material?

Page 169. How is the development of the IC (integrated circuitry) and miniaturization related to challenging "EBEs in their own territory?" We do not worry much about technical developments of Third World countries. Why would they be concerned about an inferior and fairly primitive civilization?

Page 175. You state, "Most high-ranking officers at the Pentagon and key members of their staffs knew that Roswell technology was floating." This runs counter to the statements from all of the senior officers I know. It also runs counter to statements from most of your contemporaries who went on to attain multiple stars. They specifically stated they do not have such knowledge.

Page 179. Here EBEs are poised to initiate nuclear war between the U.S. and USSR. What supports this thesis? In fact, in the instances in which UFOs interacted with U.S. missile sites, they seemed to shut down systems, not launch them.

Page 179. Lasers are not omniscient weapons. In fact, their range in the atmosphere is a major problem. At high altitudes they can reach very long distances, but not near ground level. It is not until the 1998 DoD budget that serious money is being put into the Airborne Laser Lab, which is a prototype and years away from fielding. (Note that DoD budgets are posted years in advance.)

Page 182. The nutrition comment sounds like the old *Twilight Zone* episode called "To Serve Man." Again, if ETs possess advanced technology, why would they not understand cloning and raise their own food? Domestication of animals was one of the first fundamental steps in establishing the Agricultural Age. It was the first step above hunter-gatherers. How would ETs have advanced so far without creation of a renewable food supply?

Page 184. Laser range-finders have improved accuracy. However, IFF (Identification Friend or Foe) remains a critical problem.

Page 186. No HEL (high-energy laser) was ever fielded with DSI. We are still (1997) researching how to employ high-energy lasers as weapons. Congress has just recently again provided funding for the airborne HEL test bed. It is a long way from fielding. (Still true.) At SDI Lt. Gen. Malcolm O'Neil, himself a Ph.D.-level laser physicist, never trusted DE and stressed hit-to-kill mechanisms for the years he headed the effort.

Page 196. EMP as a weapon will fry electronics and do permanent damage. However, the UFO reports claim that electronics are shut down, and later restored. That is not consistent with EMP.

Page 196. Claims that we could identify UFO signatures. To do so means that the information about those signatures must be widespread. There is no indication that operators of sophisticated electronic equipment were so informed.

Page 206. What did Nixon surrender to the Chinese? We left Vietnam and it fell to the North. They continue to skirmish with the Chinese. China has not expanded its influence save for Hong Kong, but that was a British decision.

Page 208. Who were the "dangerous academics" and what made them of concern?

Page 216. My friends at Los Alamos who knew Feynman well do not support comments that he was aware of ETs or the Roswell crash.

Page 217. Stealth did not come about via any Roswell technology. Ben Rich has discussed the evolution in his book. From our discussions I can state it was purely terrestrial in origin.

Page 218. The F-117 is not crescent-shaped. It has flat edges made for the easiest angles for complex calculations. Contrary to popular belief, "stealth" does not equal invisibility.

Page 220. There is no relationship between dU (depleted uranium) and going nuclear. This extremely heavy metal allowed us to defeat armor more effectively. The issue with war in Central Europe was the number of tanks the Soviets could present. There were just too many targets for conventional ammo, dU or otherwise. It was the tank onslaught that might prompt a nuclear response, not killing individual tanks.

Page 220. The A-10 is a "Warthog" not a "Hedgehog."

Page 232. DE (directed energy) is not synonymous with particle beam weapons. Lasers and acoustic weapons are also forms of DE.

Page 234. ARPA (now the Defense Advanced Research Projects Agency—DARPA) is not "highly secretive" though some of their projects may be. Their RFPs (requests for proposals) are listed on the CBD (Commerce Business Daily).

Page 237. Why would Gen. Trudeau be responsible for Agent Orange? In a briefing I was giving a few years ago, I had an ex-DDR&E state that he gave the okay for Agent Orange.

Page 248. Neither the Army Space Command, nor any other organization deployed particle beam weapons. Also mentioned are "missile-mounted kinetic energy beam" weapons as ASATs (anti-satellite) and for law enforcement. A KE (kinetic beam) is an oxymoron. As mentioned before, KE systems were hit-to-kill. I don't understand why DoD would give LEAs (law enforcement agencies) any missiles, let alone ASATs.

Page 252. Check Gen. Trudeau's history. I think this is wrong and does not account for his time as a corps commander.

Page 264. Is Agenda B really an AGENA B? If so, it is a rocket not a satellite.

Page 265. No satellites "swoop in from higher orbit." While the Brilliant Pebbles concept called for maneuvering satellites, they were never built or deployed. Certainly the ancient Gemini system did not have the maneuverability described.

Page 266. Describes "over fifty years, now, war against UFOs." If so, this goes well beyond your tenure and would be indicative of a massive effort and includes many, if not all, of the senior leadership of the military and political systems. Who can support this claim of yours? **Page 268.** A UFO was reportedly shot down over Ramstein (Air Force Base in Germany)? Who has it? This is near Frankfurt, a major metropolitan area and well west of the border, which would make shooting down a craft of any kind a very significant event. I have never heard of this case from either American or German sources considered to be reliable. There are some rumor-level papers on the case. Is it hard?

This is not a complete list. There are also many history of technology questions that I am attempting to get background on. However, if you could provide answers to each of these questions, it would help us understand what happened and would sell the story. I hope thinking about some of these sticky issues proves useful for both of us.

Sincerely,
John B. Alexander, Ph.D.
U.S. Army (retired)

SUBCOMMITTEE ON SPACE AND AERONAUTICS
U.S. HOUSE OF REPRESENTATIVES
COMMITTEE ON SCIENCE, SPACE,
AND TECHNOLOGY
2320 RAYBURN OFFICE BUILDING
WASHINGTON, DC 20515
(202) 225-7858
(202) 225-6415 FAX

Aerial and Related Phenomena:
Is There Reason for Concern?
XX Rayburn House Office Building
Someday, May 2000
Time
The Subcommittee on Space and Aeronautics will hold
a hearing entitled, "Aerial and Related Phenomena,"
on May xx 2000.
At time, TBA , in Room TBA

Testimony before the Committee will focus on: 1) evidence supporting or refuting observations made by credible sources of physical and intellectual phenomena that are not conveniently explained; 2) the collection of physical and sensory (including, but not limited to, photographic, radar, radiation, materials, etc.) evidence suitable for post-event scientific analysis; 3) the range of evaluations and interpretations of these events and evidence; 4) the potential for threat to national security; 5) how the U.S. Government and scientific community has responded to such evidence; 6) whether or not the U.S. Government has withheld the release of information about the topic; 7) and leading to the commencement of open scientific evaluation and debate of the veracity and implications of the information available from all sources, including that currently held and that collected in the future.

Scheduled witnesses include:

— A former military officer involved in studying the topic for the government
— An informed skeptic
— Former military officers with firsthand sightings and interference with strategic systems
— Commercial airline pilots with recent firsthand experience
— An internationally acknowledged topical expert

This was drafted in December 1999 and was never used. A possible witness list was also drafted. That is withheld as some of the people I suggested as witnesses were not aware of the potential for hearings and did not know that they might be called upon.

BACKGROUND MATERIAL FOR HEARINGS ON AERIAL PHENOMENA

Invariably there is much controversy associated with the study of any phenomenon. There are mercurial evidentiary aspects, strongly held belief structures, and unyielding precedence that warrant trepidation. As will be demonstrated, there are incidents of unknown origin yet undeniable reality. They make open examination of these events not only reasonable but issues of national security and air safety, coupled with documented public concern bodes them obligatory.

Over thirty years ago the Condon study delivered a report that attempted to convince the public that unidentified flying objects (UFOs) could be dismissed scientifically and that they posed no threat to the security of the United States. Today it appears that the committee's report was either fundamentally flawed or purposefully deceptive. The report was correct in that there has been no invasion from outer space. But, as the evidence will establish, that does not mean the events are not real and, in fact, they may pose a threat to national security. Additionally, UFOs repeatedly have been observed in our commercial airspace and even interfered with navigational equipment on flights.

It will not be our intent to reexamine the prior report other than to point out that the basic document was internally inconsistent in that the conclusions were directly refuted by the evidence in the body of the report. The very skeptical committee admitted that 6 percent of the cases evaluated remained unidentified. Still true today, the testimony that will be presented will demonstrate that UFO sightings and related phenomena continue. Further, it will be shown that there is both hard evidence and credible witnesses to support the reality of these events.

The real damage of the Condon Committee report was that it not only summarily dismissed all UFO reports, but created an atmosphere in which reporting or studying such events was treated with abject scorn. For a person to have any involvement with UFOs, whether voluntary or involuntary, often meant being subjected to intense personal ridicule. It was because of this artificially generated hostile atmosphere that many veridical cases have gone unreported and virtually no research has been done by scientific or academic institutions of any standing.

Despite official condemnation, polls consistently show that a majority of the public believes that UFOs are real. A cavernous gulf exits between the opinions of the general public and those of traditional scientists. The reason for this disparity is an absolute reductionist approach by most scientists who, in general, want repeatable experiments, and the public, 7 percent of whom report having seen a UFO and have more open beliefs. They trust the testimony of friends, relatives, and others. To quote one scientist who is willing to explore the topic, "If I can't trust my mother, who can I trust?" To be sure, many of the reported incidents are misidentifications of objects and have prosaic explanations. The most difficult problem for researchers is discrimination between veridical and mendacious reports. There remains, however, a core set of observations and events that defy all known theories. It is upon such cases that we shall concentrate in these hearings.

It is important to note that from the beginning, the government's history in reporting accurate UFO information to the public has been very poor. There have been several reasons for this institutional behavior. In the 1950s there was concern that confirmation of UFOs of extraterrestrial origin might lead to a breakdown in social order. Also, the CIA used UFO sightings as a cover story for inadvertent sightings of the high-altitude and highly classified U-2 spy plane. The U.S. Air Force has concocted several stories to explain what has become world renowned as the Roswell event. They attempted to ignore, then stifled congressional inquiry by the late Congressman Steve Schiff when he asked for an explanation that he could provide to his constituents in New Mexico. To this day the answers provided to the public do not fit the facts of this dramatic case. Further, several agencies have been found withholding FOIA requested UFO documents long after they stated they had no more left. These evasive actions and failure to take seriously the public's requests for information have widened the credibility gap that exists between the government and the people, and have served in a small way to undermine the fiber of democracy.

In general, the evidence supporting UFOs is based on transient events that do not lend themselves to laboratory studies. Still, there exists a body of multisensory data consisting of photographs, videotapes, movies, radar reports, space-based sensor systems, radiation readings, personal injuries, electrical interference, landing traces, and even anomalous material samples. Additionally, in cases of unexplained phenomena that may be related, there is more tangible evidence such as mutilated animal bodies and complex designs in crops. This evidence is often dismissed quickly by most scientists who fail to take the time to learn all of the facts surrounding the events. Simply put, satanic cultists are not cutting up cows and pranksters cannot account for the creation of expansive and intricate patterns in relatively brief periods. Therefore, contrary to the belief of most scientists, hard evidence exists and is waiting to be evaluated.

There is a popular misnomer that UFO sightings are mostly fleeting glimpses of luminescent objects high in the sky. Skeptics ubiquitously refer to the misidentification of the brightly shining planet Venus as the most likely explanation for UFOs. The reality is that many of the sightings take place in broad daylight. Some occur at close range—a few yards or even feet—and on rare occasions, may even include physical contact. As will be shown, in some cases events do recur—sometimes for relatively long periods of time, even years.

There are recent cases of importance. On October 26, 1999, pilots from two major airlines reported seeing a large luminescent object over the Dallas, Texas, area. When reported, the FAA indicated they had neither contact with the craft nor was it on their radar screens. On August 9, 1997, Swissair Flight 127 reported a near miss with a UFO. It was so close the pilot reported instinctively ducking. On May 25, 1995, an America West pilot reported an unidentified 300- to 500-foot object flying lower than his aircraft. Again, there was no FAA radar contact. More ominously, on March 12, 1977, the autopilot of a United flight was perturbed by a UFO taking the plane off course. The FAA in Boston noticed the flight going astray but had no contact with the UFO seen by the pilot.

Two of the most striking military cases occurred continents apart in late December 1980. One involved close contact with U.S. personnel stationed at a NATO air base located in the United Kingdom. The second case, which happened near Houston, Texas, led to the serious injury of two women and the traumatizing of a small boy. As military helicopters were identified in close proximity to the UFO, this case yielded a lawsuit against the U.S. Government.

The Royal Air Force and the U.S. Air Force jointly operated bases collocated at Woodbridge and Bentwaters. It is known that the United States stored special weapons there, thus security was very rigorous and security personnel extremely well trained. On two nights during the Christmas season of 1980, luminescent objects were observed near the weapons storage area and below tree level. Among the observers of the second incident was then Lieutenant Colonel Charles Halt, the U.S. deputy base commander. Lieutenant Colonel Halt took radiation readings that proved to be above normal and saw a glowing object maneuver through the woods a short distance away from him. The object suddenly split into five separate pieces and shot off in different directions. Another object came overhead, stopped, and shined a beam of light at his feet. He reports that a senior air policeman may have actually touched the craft on the prior incident. Despite possible interaction between these unidentified objects and the weapons storage area, no tests were made of the strategic systems stored there. It was determined that if they couldn't explain the incident, it was better to ignore it than face pandemic ridicule.

On December 29, 1980, Betty Cash, Vickie Landrum, and Landrum's seven-year-old grandson, Colby, were driving on a narrow deserted country road about forty miles from Houston, Texas, when they encountered an unknown object with flames shooting out the bottom. They stopped the car and got out to observe the rhombic-shaped craft as it hovered tenuously a little above the road. In addition to the orange-colored object, they noted a number of identifiable military helicopters following the apparently disabled craft. The craft soon disappeared and the women continued home, albeit very frightened. The following day they were quite sick and medical examination confirmed high-dosage radiation poisoning. Assuming some form of government experimental craft had injured them, they sued for damages. Extensive searches by U.S. Army representatives failed to locate the helicopters although other credible witnesses confirmed seeing those choppers two hours later. Despite the research done in support of the civilians and government confirmation that they had been seriously injured, their lawsuit was denied.

Unexplained incidents also occurred within our strategic defense systems in the United States. During a period from October 1975 to January 1976 sightings by NORAD installations were so prevalent that they became known as the Northern Tier Sightings. Missile facilities from Malmstrom Air Force Base, Montana, to Minot Air Force Base, North Dakota, to Wurtsmith Air Force Base, Michigan, to Loring Air Force Base, Maine, reported sightings of unknown lights and craft intruding in their areas. Malmstrom Air Force Base, on November 7 reported a football field–sized craft that illuminated an ICBM silo. When approached by an interceptor, the craft shot straight up and was tracked on radar to 200,000 feet. A decade earlier, on March 16, 1967, multiple ICBM control systems were inexplicably shut down during UFO incidents.

Despite thorough examination by contractors, no reason for the interference could be determined. Further, other missile silo operators were not informed about the potential problems.

The United States was not alone in experiencing problems with UFO-related interference with strategic systems. On October 4, 1982, a 900-foot in diameter UFO was seen by hundreds of people as it hovered for hours near a Soviet missile base at Byelokoroviche. During that period of time an unknown force took over the launch operations and began to prepare the missiles for a strike against the United States. This terrified the launch control officers as they were unable to stop the missiles and were gravely concerned that the United States would retaliate believing the Soviets had initiated a war. As suddenly as it began, the system shut itself down. Investigators from Moscow under direction of Colonel Boris Sokolov totally disassembled the site but failed to find a rational explanation—such as electronic malfunction—for the event. Numerous such sightings led General Igor Maltsev, Chief of Staff of the Soviet Air Defense Forces, to publicly state, "Skeptics and believers both can take this as official confirmation of the existence of UFOs." Contextually this is an astonishing statement as his predecessor was fired when a Cessna landed near Red Square.

Tactical weapon systems have also failed when encountering UFOs. A well-researched case occurred north of Tehran, Iran, on September 19, 1976. An extremely bright object was spotted from Mehrabad control tower. Two American-made F-4 fighters were scrambled to investigate and could see the object for a distance of 70 miles. Both airborne and ground radars confirmed the object. As the first F-4 closed to within 25 nautical miles it lost all instrumentation and communications. After the pilot broke off the intercept and turned away, the aircraft regained all of its capabilities. The second F-4 pilot noted a smaller object coming at him at high speed. At the instant he attempted to fire an Aim-9 missile at it, the F-4 lost power to the weapons control panel as well as its communication systems. This case involves multiple witnesses at different locations, radar verification, electrical interference on the aircraft, failure of a weapons system, and reports of a highly maneuverable unexplained craft.

From October 1989 until April 1990 there were hundreds of reports of UFOs over Belgium. The intrusions were so pervasive that Colonel Wilford De Brouwer (later Major General and Deputy Chief of Staff of the Belgium Air Force) created a special task force to investigate the sightings. Of special note was a pursuit by a Belgian F-16 on the night of March 30 and 31. Visual contact was made followed by a radar lock. Each time the radar locked onto the UFO, the target would rapidly change speed and altitude and evade pursuit.

Concerned about these flights, they inquired if they could be American experimental test vehicles. They received assurances they were not.

Starting in 1987, Gulf Breeze, Florida, became host to many UFO sightings that recurred for several years. They were sufficiently prevalent that groups of people gathered nearly every night at the inlet shore. Many people photographed these objects. The witnesses of these events number in the thousands. The best photographs have been subjected to extensive investigation and widely published.

On March 13, 1997, undeniable sightings took place over Phoenix, Arizona, and were captured on several video cameras and witnessed by thousands. For reasons unknown, this event did not blossom in the national media until June. While focus was placed on the Phoenix Lights, the reality was the event was far more extensive. It began near Henderson, Nevada. Consistent time-sequenced sightings were made across northern Arizona, videotaped near Prescott, and later seen in Phoenix. From Phoenix the very large V-shaped craft of unknown origin ventured farther south and was located as far away as Tucson before a second sighting was made over Phoenix. No viable explanation has been forthcoming from any source.

In addition to sightings of UFOs, there is a plethora of reports from individuals who believe they have had encounters with extraterrestrial beings. These experiences have been inculcated into every part of our society and now frequently appear in popular advertisements. The origin of these events is highly controversial. Regardless of whether or not one believed in the existence of aliens or the etiology of the events, there are sufficient numbers of people reporting these incidents so as to make it imperative that the topic be seriously studied.

Also on the research agenda should be the study of the societal impact of an officially confirmed announcement of the existence of intelligent extraterrestrial life. Whether one assesses the probability as extremely high or vanishingly low, the impact is likely to be highly significant. Private groups have conducted some exploratory work. However, even low probability of occurrence dictates preparedness.

Government secrecy has played a pivotal role in UFO mythology. Of course some secrecy is necessary and appropriate. However, when declassification of FOIA documents is absurdly conducted, the public is left to fill in the blanks with their active imaginations. One UFO researcher proudly displays documents he received under FOIA in which the pages are almost totally blacked out. Such actions tend to support the notion that information about UFOs is known to the government but suppressed from the public. The stories that have been created are epidemic and retold in books, movies, and television programs and their ratings are always high. Episodes, all of which have serious advocates, run from the preposterous—the government has a secret treaty and trades technology—to the improbable—we are reverse engineering crashed saucers—to the more likely—that the issue has been ignored despite continued incidents. Today there are many former government employees who believe that if they say anything about UFOs they are likely to lose their pensions and be placed in jail. There is an exigent need for public clarification of the government's classification policy regarding UFOs. That pronouncement alone would resolve many fanciful tales.

Funding of UFO research has been extremely scarce. In recent years in the United States it has been limited primarily to three philanthropic endeavors. After several years of sponsoring independent projects, in 1995 Robert Bigelow created the National Institute for Discovery Science. Joe Firmage developed the International Space Science Organization in July 1999. And Laurance Rockefeller has provided funds to a variety of small projects. Most other research efforts have been conducted out of the pockets of people with modest means. Regardless of whether or not substantial private money could be made available, there is no matching the national collection capabilities of governments.

The time is long overdue to erase the Condon stigma that has suppressed most efforts to report, or scientifically investigate, these phenomena. As a bare minimum it is essential that researchers be free to explore anomalies without fear of retaliation and intimidation by peers and media. The government should be more forthcoming about the information gathered by various agencies. Further, agencies should be encouraged to fund objective research projects. Whenever possible they should release information about anomalies that is routinely collected to researchers attempting to solve these complex problems.

In conclusion, despite fifty years of official denial, UFO events continue to occur. Those cases presented here are only a representative historical sample. These events still happen to credible people and provide physical evidence. They happen in daylight as well as nighttime. Some observations are made at close range in which misidentification can probably be eliminated. This is a global issue, one that demands concerted efforts at resolution. Given the multifaceted presentations of the phenomenon it is unlikely that a single simple answer will be found. Rather, these anomalous events represent enigmas. They evoke the essence of human curiosity in our indomitable and unending quest for a technological explanation of our true place in the universe.

HOUSE SUBCOMMITTEE ON SPACE & AERONAUTICS

Aerial and Related Phenomena:
Is There Reason for Concern?

*Opening Statement of Chairman
Someday*

For several years many of our constituents have been asking us to conduct hearings on the topic of aerial phenomena and related events. Though some people believe this topic was resolved decades ago, there continue to be sightings of unexplained objects by highly credible people. While some of these reports may be misidentification of natural objects, other sightings defy all attempts at rationalization. The simple solutions postulated decades ago appear to many to be invalid. As will be demonstrated here today, there are incidents of unknown origin, yet undeniable reality, that make open examination of these events not only reasonable, but issues of national security coupled with documented public concern bode them obligatory.

Invariably there is much controversy associated with the study of any phenomenon. There are mercurial evidentiary aspects, strongly held belief structures, and unyielding precedence that warrant great trepidation. A Byzantine mythology has enveloped the topic of aerial phenomena, obfuscating reason, suffocating truth, and permanently branding all those who come in contact, whether willingly or unwillingly. Too frequently polar positions are established, often immutable yet devoid of fact. This committee is not exempt from the possible stigma associated with the topic. Therefore I want to thank each of our committee members for having the intestinal fortitude to come, to listen carefully, and to participate in these hearings.

In general, the scientific community rejects the notion that the extreme distances could be traversed to arrive near Earth from another galaxy. Those judgments are based on their current understanding of laws of physics. If one but contemplates the tremendous intellectual advances of the last century, you realize how vainglorious it is to believe that we have obtained the ultimate knowledge of the boundaries of time and space. Scholars have invariably defined impenetrable lines of demarcation only to have them tested and defeated. Given our current multibillion year estimates of the age of the universe, is it not reasonable that somewhere in the vastness of space there exists one or more advanced civilizations that have developed both the knowledge and engineering to transit intergalactic distances? It is foolish to believe that because we do not yet know how to do this that it cannot ever be accomplished under any circumstances. The question that is before us is whether or not there is any veridical evidence that supports the sightings of various aerial phenomena that are being reported, and if so, does that evidence constitute grounds for further scientific research and investigation?

If craft have indeed been able to cross these boundaries and are interacting with us in close proximity to Earth, that would answer the Aeolian question and prove that we are not alone in the universe. How profound that would be. If true the implications are astounding and could potentially permeate every aspect of our lives. If not, the study of these events is likely to lead to a deeper understanding of our perceptions of the universe.

On a less grandiose scale, there are repeated reports of interaction with and occasionally interdiction of some of our most sophisticated defense systems. Even a prosaic answer to those events has profound consequences and demands investigation. The reality of these reports is undeniable. It is unconscionable that the specter of calumniation has prevented rigorous research into these events.

Conjecture abounds concerning the government's role is suppressing information about this topic. The public assumes that given the pervasiveness of the advanced sensor systems in our national collection means that it is inevitable that observations must be made routinely. Therefore, it seems reasonable to assume that a substantial body of data must exist somewhere within those agencies. Vociferous denials about retention of information by government agencies have too often later proven false. This has led to a general distrust between researchers and the government. Frequently, withholding information has been cloaked in national security concerns. A simple statement of what the classification policy is regarding aerial phenomena would alleviate most of that problem.

Today we will listen to those who have studied aerial phenomena extensively and those who have experienced it firsthand. It is time to listen carefully to the facts and determine if we as a nation are prepared to accept responsibility for supporting investigation of this complex problem that may ultimately address the nature of our place in the universe.

John Alexander

September 13, 1999

INDEX

abductions, 231–32, 248–49, 255
   military, 256–63
Abrahamson, Jim, 6, 33, 34, 35, 39, 260
Advanced Theoretical Physics Project (ATP),
   6–39, 57, 66, 72, 76, 105, 139, 175, 178,
   180, 215, 223–24
Aftergood, Steven, 64
Air Force (USAF), 31–32, 33, 38, 69, 86, 103,
   113, 124, 131, 240
   Condon Report and, 51–57, 64, 82, 87, 96, 192
   JANAP-146, 96–97
   Phoenix Lights and, 86
   Roswell and, 192, 243, 288
   Science Advisory Board of, 147
   space surveillance system of, 210
Akdogan, Haktan, 179–80
Aldrin, Edwin "Buzz," 139–40
Alexander, Victoria, 242–43, 257, 258, 262
Allen, Paul, 26
Allin, Mark, 206
Ames, Aldrich, 207
Anderson, Jack, 8–9
Antarctica, 180–81
Apollo program, 133, 134–43, 224, 248
Area 51, 14–15, 21, 41, 88, 89, 144, 146, 151,
   162, 203, 255
Armstrong, Neil, 134, 136, 139
Army Science Board (ASB), 36–38
Arnold, Ken, 2
Aviary, 127–28
Azizkhani, Mohammad Reza, 197

Backster, Cleve, 12
Balducci, Corrado, 242

Bartley, James, 262, 263
Bassett, Steve, 83–84, 108
Beckwith, Charlie, 9
Belgium, 215–16, 225–26, 251,
   291–92
Bentwaters incident, 155–60, 189–90, 200,
   208, 211, 237, 289–90
Berkner, Lloyd V., 123
Berliner, Don, 109, 126
Bernstein, Carl, 125
Bigelow, Robert, 41, 42, 66, 72, 93, 232, 233,
   235, 249, 293
Birnes, Bill, 244–45
Black Book, 204
Blackburn, Ron, 14–15, 21
Black World, 80, 84, 208
Blanchard, William, 246–47
Blue Book, 19, 54, 55, 56, 86, 175–76, 189,
   202, 203, 244
Blum, Howard, 37
Braddock, Joe, 16
Bradlee, Ben, 125
Branson, Richard, 26
Brazil, 178, 240
Brennan, John, 30
Bronk, Detlev, 123
Buran, Fred, 157
Burns, Bill, 41
Burns, Frank, 10
Burroughs, John, 157
Buser, Rudy, 47
Bush, George W., 114, 204
Bush, Vannevar, 123
Bustani, Jose Mauricio, 178

Byelokoroviche/Khmelitskiy incident, 170–72, 291

Cabansag, Edward, 157
Cameron, Lou, 47
Canada, 214, 216–18, 220
Canavan, Gregg, 24
Carter, Jimmy, 44, 101–5, 114
Cash, Betty, 160–61, 289, 290
Cash-Landrum incident, 25, 78, 160–62, 236, 289, 290
cattle mutilations, 139, 234–35, 250–51, 263, 289
cell phones, 196–97
Central Intelligence Agency (CIA), 29–30, 37, 39, 53, 57, 69, 131, 207, 240, 288
  FOIA and, 99–100
  Majestic 12 and, 124, 127
  Robertson Panel of, 54–55, 56, 124
Chandler, J. D., 157
Channon, Jim, 9–10
Chernovsky, Igor, 172
Chile, 179, 240
China, 178–80
Christianity, 227, 242
Churchill, Winston, 188
Clancy, Tom, 38–39, 139–40
Clemmons, Albert, 213–14
Clinton, Bill, 101, 108–11
Coleman, Bill, 200–203, 207, 208
Collins, Bob, 91–92, 127
COMETA Report, 218–19
Committee for Skeptical Inquiry (CSI), 118, 119, 224
Communications Instructions for Reporting Vital Intelligence Sightings (CIRVIS), 96–97
Condign Report, 192–93
Condon, Edward U., 51, 56–59, 61
Condon Report, 19, 51–64, 82, 87, 96, 102–5, 113, 118, 121, 141, 175, 189, 192, 287–88, 293
Congress, 65–74, 82–84
conspiracy theory, 1, 4, 17, 39, 74, 80, 96, 101, 113, 115, 122, 204, 206, 269
  Apollo program and, 136
  Blue Book and, 175
  cattle mutilations and, 250–51
  Clinton and, 108, 110

defense contractors and, 93–95
Eisenhower and, 106–7
Majestic 12 and, *see* Majestic 12
Reagan and, 112
reverse-engineered UFOs and, *see* reverse-engineered UFOs
secrecy and, 196, 203
Continuity of Government (COG), 130–32
Cooper, Gordon, 202–3, 220
Cooper, Tim, 126
Corso, Phillip, 40–50, 203, 218
  Alexander's letter to, 41, 275–84
Coyne, Lawrence, 162
crashes, UFO, 144–45, 222–23
  at Roswell, *see* Roswell incident
crop circles, 185, 193–94, 289
Cuba, 135
Cutler, Robert, 128
Cutler-Twining Memo, 128–30

Dallas, Texas, incident, 172–73
Davis, Tom, 73
Dean, Robert O., 212–14, 217
De Brouwer, Wilfred, 212, 215–16, 220, 225–26, 291
defense contractors, 93–95, 217
Defense Intelligence Agency (DIA), 28–29, 37, 39, 91
Department of Defense (DoD), 6, 10, 35, 52, 73, 79–82, 89–90, 94–96, 100, 103, 109, 130–31, 147, 271–72
Department of Energy (DoE), 81, 89–90, 109
Disclosure Project, 73, 146–47
Downie, Leonard, Jr., 125
Dulce, New Mexico, 263–65, 266

Edwards Air Force Base, 106, 107
Einstein, Albert, 1, 117, 118, 144, 255
Eisenhower, Dwight D., 43, 105–7, 123, 130, 131, 136, 188
Eisenhower, John S. D., 107
Eisenhower, Laura Magdalene, 107
Eskimo Scouts, 163–65
evolution, 242
extraterrestrial encounters, 232, 255, 268, 292
  abductions, 231–32, 248–49, 255, 256–63
  Apollo program and, 136–43, 248
  confirmation of, 240–43, 292
  Corso's account of, 47–48

Dean's account of, 213, 214
international space law and, 142–43
wars, 266–67

F-117, 14, 22, 88, 165, 216
Federal Aviation Administration (FAA), 73,
    92–93, 289
Federal Bureau of Investigation (FBI), 124, 207
Felt, Mark, 124–25
Fermi, Enrico, 90
Feynman, Richard, 228, 229–30, 232
Ford, Gerald, 112–13
Forrestal, James, 123, 131
Fort Dix, 97
France, 187, 218–19, 220, 239–40
Frazier, Ken, 224, 225
Freedom of Information Act (FOIA), 16, 29, 30,
    78, 98–100, 128–30, 187, 205, 288, 292
Friedman, Stanton, 99, 128, 132
Friend, Bob, 243–44, 249
Frosch, Robert, 103–4
Fulford, Earl, 244–45

Galbraith, Marie "Bootsie," 109
Gamboa, Pierantoni, 178
Gemini photographs, 202–3
General Accounting Office (GAO), 74, 82,
    243, 244
Germany, 180
Gevaerd, A. J., 178
Gibbons, John, 108–9
Gilliland, James, 2
Global Hawk, 209
Global War on Terror (GWOT), 79, 94, 96,
    194, 204, 209, 217
Good, Timothy, 123, 147
Gorbachev, Mikhail, 155, 239
Gore, Al, 114
government, 75–100, 272
    budgets and priorities in, 79–84, 95–96,
        100, 270
    bureaucracy in, 78–79, 82–83, 87, 203–4
    cattle mutilations and, 250–51
    chain of command in, 80
    change resisted in, 77–78
    Congress, 65–74, 82–84
    defense contractors and, 93–95
    Executive Branch of, 78, 80–81, 85, 113–14,
        131, 272

ineffectiveness of, 98, 100
Judiciary Branch of, 78
Legislative Branch of, 78, 80
personal interest vs. responsibility of,
    85–87, 100
presidents, *see* presidents
secrecy and, *see* secrecy
UFOs seen by employees of, 85–87, 100
see also specific agencies
Gray, Gordon, 123, 124
Gray, L. Patrick, 124
Greenbrier, 132
Greer, Steven, 73, 146–47
Groom Lake, 21, 89, 144
Gulf Breeze, Florida, incident, 166–68, 292

Haines, Gerald, 30, 127
Haines, Richard, 36–37, 73, 172
Halt, Chuck, 157–60, 208, 211, 290
Hanssen, Robert, 207
Hastings, Robert, 199–200, 272
Haut, Walter, 246–47
Have Blue, 88
Hawking, Stephen, 230–32
Hellyer, Paul, 116, 216–18
Henry, Richard, 104–5
Hillenkoetter, Roscoe H., 123, 124
Hill-Norton, Lord, 183
hoaxes, 226, 271
Holloman Air Force Base, 106–7
Hopkins, Budd, 248
Horgan, John, 64
House Subcommittee on Space & Aeronautics,
    294–96
Hubbell, Walt, 101
Hubbell, Webster "Webb," 108
Hunees, Antonio, 109
Hunsaker, Jerome, 123, 124
Hyman, Ray, 119
Hynek, J. Allen, 55, 63, 104, 113, 175, 220
hypersonic flight, 147

IHEDN, 218–19, 220
infrared sensors, 46–47
Intelligence Community (IC), 17–18, 37, 98,
    124, 131, 196–97
Iran, 18, 197–98, 225–26, 291
IT 6, 91
Ivy Bells, 206–7

Jacobs, David, 248
Jafari, Parviz, 197–98, 225–26
JANAP-146, 96–97
Jezzi, Arrigo, 162
Johnson, Quint, 92–93
Jones, C. B. "Scott," 91, 105, 127

Katina, Don, 32
Kean, Leslie, 110
Kelleher, Colm, 93, 232, 235, 264
Kelly, Tom, 10
Kennedy, John F., 114, 128, 134–36
Khrushchev, Nikita, 135
Klass, Phil, 104, 163, 190, 198, 252
Knapp, George, 41, 170–71, 172, 232, 235,
    263, 264
Kucinich, Dennis, 66–67
Kupperman, Bob, 141–42
Kupperman, Helen, 142–43

LaBerge, Walt, 35–36
Laird, Melvin, 109, 110
Lakenheath, 189–90, 193
Lammer, Helmut, 256–62
Landrum, Colby, 160, 289, 290
Landrum, Vicky, 160–61, 289, 290
La Plume, Steve, 158
Lazar, Bob, 41, 89, 207–8
Le Carré, John, 122
Lenker, Karl, 172–73
Levine, Norman, 57
Lockheed Skunk Works, 14, 21–23, 39, 88,
    167–68
Long Endurance Multi-INT vehicles
    (LEMVs), 209
Los Alamos National Laboratory (LANL),
    21, 22, 25, 29, 38, 89–92, 168, 184, 214,
    228, 265–66
Lovekin, Stephen, 73
Low, Robert, 57

Maccabee, Bruce, 30, 167, 168, 248
Mack, John, 248–49
Majestic 12 (MJ-12), 4, 91, 121, 122–33
    Aviary and, 127–28
    beginnings of, 122–24
    CIA and, 124, 127
    Continuity of Government and,
        130–32

*Extraterrestrial Entities and Technology,
    Recovery and Disposal* manual, 126–27
    National Archives and, 128–30, 132
    Watergate compared with, 124–26
Malmstrom Air Force Base, 168–70, 200, 208,
    237, 290
Maltsev, Igor, 173–74, 291
Mansfield, Ohio, incident, 162–63
Manhattan Project, 117–18, 135, 228
Mantell, Thomas, 213–14
Marcel, Jesse, 244–48
Marcel, Jesse, Jr., 244–45, 247–48
Massachusetts Institute of Technology
    (MIT), 249
Maussan, Jaime, 180
McCain, John, 86–87, 115
McCaslin, Patrick, 175–76
McConnell, Howell, 30–31, 33, 34, 205, 208
McCrae, Ron, 8
McDonald, James, 59–60, 63
McGaha, James, 225
McGuire Air Force Base, 97
Meiwald, Fred, 170
Menzel, Donald, 123, 124, 132
Mexico, 180, 240
MIAs, 42
Miller, Norman C., 111
mind control, 256
missiles, 168–72, 225, 237–38, 290–91
Mitchell, Edgar, 73, 138, 139, 140, 248
Montague, Robert M., 123
Moore, Bill, 91, 123, 126, 127
Moore, Robert, 13, 20, 32
Morrison, David, 140–41
Moynihan, Daniel Patrick, 204
MQ-1 Predator and MQ-9 Reaper, 209

National Academy of Sciences, 118
National Aeronautics and Space
    Administration (NASA), 102–5
    Apollo program, 133, 134–43, 224, 248
    budget of, 135
National Archives, 86, 240
    Majestic 12 and, 128–30, 132
National Institute for Discovery Science
    (NIDS), 41, 93, 138, 140, 232, 233, 235,
    249, 250, 251, 263, 293
National Research Council (NRC), 119
National Security Act, 131

National Security Agency (NSA), 30–31, 33, 34, 36, 37, 39, 87
National Security Council, 131
Nellis Air Force Base Test Range, 145–46
Niemtzow, Richard, 162
night vision, 46–47
Nixon, Richard, 109, 114, 124
Noll, Melvern, 251
Noory, George, 145
North American Air Defense Command (NORAD), 26–27, 32, 225, 290
North Atlantic Treaty Organization (NATO), 212–21, 251
  Assessment document and, 212–14
  Canada and, 216–18, 220
  De Brouwer account and, 215–16, 220
  France and, 218–19, 220
  SHAPE and, 212, 213, 214
Northern Tier sightings, 59, 290
nuclear weapons:
  development of, 117–18, 206, 245
  lost, 244

Obama, Barack, 79, 115, 117, 146, 204
Oberg, Jim, 203
Odom, William, 13
Office of Science and Technology Policy (OSTP), 102–4, 108–9
Onet, George, 250
Oppenheimer, J. Robert, 269
Oxburgh, Ronald, 183–85, 186, 193

Page, Thornton, 54, 60, 61
Paynter, Bill, 111
Pell, Claiborne, 68, 105, 119
Pelton, Ronald, 207
Penniston, Jim, 156–57, 159, 208, 225
Perrizo, Jerry, 91
Personnel Reliability Program (PRP), 155, 169, 175
Peru, 179
Pfautz, Jim, 37
Phoenix Lights, 86, 165–66, 210, 225, 253, 292
Plantonev, Vladimir, 171
Podesta, John, 109–10
Pollard, Jonathan, 207
Pope, Geoffrey, 186–87
Pope, Nick, 183, 185–87, 189–92, 194, 199
Powell, Colin, 10, 38

POWs, 42–43, 261
precognitive sentient phenomena (PSP), 227–28, 232, 235–38, 248
  Step Back proposal for, 238, 272–73
presidents, 80–81, 101–21, 270
  briefing of, 116–21
  Carter, 44, 101–5, 114
  Clinton, 101, 108–11
  Continuity of Government and, 130–32
  Eisenhower, 43, 105–7, 123, 130, 131, 136, 188
  election of, 114–15
  Ford, 112–13
  Obama, 79, 115, 117, 146, 204
  Reagan, 33, 44, 68, 83, 101, 110–12, 114, 239
  Truman, 23, 101, 123, 131, 239
Press, Frank, 102–4
Pritchard, Dave, 249
Project Blue Book, 19, 54, 55, 56, 86, 175–76, 189, 202, 203, 244
Project Grudge, 54
Project Mogul, 109, 244–45, 247
Project Sign, 53–54
Purcell, Francis A., 106
Puthoff, Hal, 24, 41, 42, 91, 107, 127, 131, 226

Quigley, Joan, 68
Quintanilla, Hector, 176

Ramey, Roger, 247
Randi, James, 227
Ray, Norman, 214–15
RB-47, 58
Reagan, Nancy, 68, 111
Reagan, Ronald, 33, 44, 68, 83, 101, 110–12, 114, 239
Regan, Don, 68
religion, 227, 241–42
remotely piloted vehicles (RPVs), 209
Rendlesham Forest, 157, 192, 193
reverse-engineered UFOs, 26, 115–16, 143, 144–54, 222–23, 255, 293
  energy technology and, 148–54, 224
Rich, Ben, 21–23, 39, 88
Ridpath, Ian, 159
Rivers, Mendel, 112
Roberts, Neil, 209
Robertson Panel, 54–55, 56, 124
Rockefeller, Laurance, 72, 108–9, 116, 118, 293

Rockwell, Ted, 105
Rodriguez, Jose, 172–73
Roosevelt, Franklin D., 117–18, 131
Roswell incident, 14–15, 23, 24–25, 38–39, 53, 82, 140, 145, 219, 243–48, 271
    Air Force and, 192, 243, 288
    Apollo program and, 138, 248
    Clinton on, 108
    The Day After Roswell (Corso), 40–50
    Haut and, 246–47
    Majestic 12 and, 126, 128, 131
    material recovered from, 15, 144, 151, 228, 244–48, 249
    Project Mogul and, 109, 244–45, 247
    The Roswell Incident, 123
    Roswell Report: Case Closed, 109
Rothenstein, Lou, 177
Ruano, Cervantes, 180
Rumsfeld, Donald, 204, 217
Runyon, Brad, 175, 176
Rust, Mathias, 174
Rutan, Burt, 25–26, 39, 139, 148

S-4, 89, 151
Sachs, Alexander, 117–18
Sagan, Carl, 60–63, 90, 103, 113
Salas, Robert, 73, 168–70, 208
Salla, Michael, 107
Sammet, George, 44–45
Santa Maria Huertas, Oscar, 179
Sarran, George, 161, 162
Saunders, David, 57, 63
Schiff, Steve, 82, 288
Schmitt, Harrison "Jack," 138–39, 140
Schuessler, John, 160, 161
secrecy, 87–89, 100, 113, 196–211, 271, 292–93
    classification system and, 203–5, 206, 293
    Coleman and, 200–203
    disclosure advocates and, 73, 239–40, 242, 254, 272
    document storage and destruction, 205–6
    psychological impact of, 198–200
    rumors and, 206, 207–8
    spying and, 206–7
    technological advances and, 208–10
Scully, Jay, 32, 36
Senior Trend, 88
Sensenbrenner, James, 72, 74

Shandera, Jamie, 122–23, 126–28
Sheaffer, Robert, 224
Shepard, Alan, 135
Sherman, Dan, 262
Simon, Stephanie, 252
Skantz, Larry, 31, 32
skeptics, 224–27, 253, 254, 267, 268, 270–71, 273, 289
Skinwalker Ranch, 232–35, 237
Smith, Walter B., 123, 124
Smith, Wilbert, 217–18
Society for Scientific Exploration (SSE), 62, 72, 218
Sokolov, Boris, 171, 172, 291
Sokolov, Sergei, 174
Souers, Sidney, 123, 124
Soviet Union, 8, 11, 12, 14, 25, 27, 33–34, 45, 46, 53, 135
    Alaska and, 163–64
    Majestic 12 and, 127
    NATO and, 214
    Project Mogul and, 245, 247
    Roswell incident and, 25, 245, 247
    in space race, 134–35, 136
    spying and, 206–7
    UFO sightings in, 168, 170–72, 173–75, 225–26, 237–38, 240, 291
space law, 142–43
space program, 134–36, 243
    Apollo, 133, 134–43, 224, 248
space surveillance system, 210
SR-71, 22, 23, 39
Star Gate, 9, 36
stealth technology, 206, 223
Stilwell, Richard G., 10–11
Stone, Jeremy, 68
Strategic Defense Initiative (SDI), 6, 20, 33, 34, 35, 39, 112, 260
Streiber, Whitley, 248
Stubblebine, Albert "Bert," 12, 16, 28, 29, 33, 37
Sturrock, Peter, 60–61, 72, 73, 108, 162
Subcommittee on Space and Aeronautics, 285–86
Sumner, Gordon, 44
Supreme Headquarters Allied Powers Europe (SHAPE), 212, 213, 214
Symington, Fife, 166
Szilard, Leo, 117

TA-33, 91–92

Taos Hum, 265–66

Task Force Delta, 9–10, 12, 13

Teller, Edward, 23–25, 39, 89, 90, 107, 117, 196

Thayer, Gordon, 189

Thompson, Ed, 9

Thompson, Richard, 20–21, 31

Thurman, Maxwell R., 11, 12, 35

Thurmond, Strom, 49

Torres, Milton, 190–92, 193, 195, 198–200, 211, 213, 225–26, 240

Townsend, Jack, 245

Trefry, Richard, 8, 11

Tretyak, Ivan, 174, 226

Trudeau, Arthur, 43–44, 45, 47, 136

Truman, Harry S., 23, 101, 123, 131, 239

Turkey, 179–80

Twining, Nathan F., 52–53, 123
  Cutler-Twining Memo, 128–30

Tyler, Paul, 162, 180–81

U-2, 21–22, 208, 288

UFO cases, strong, 155–82, 269
  airline pilot observations, 172–77, 289
  in Antarctica, 180–81
  Bentwaters, 155–60, 189–90, 200, 208, 211, 237, 289–90
  Blue Book and, 19, 54, 55, 56, 86, 175–76, 189, 202, 203
  in Brazil, 178, 240
  Byelokoroviche/Khmelitskiy, 170–72, 291
  Cash-Landrum, 25, 78, 160–62, 236, 289, 290
  in Chile, 179, 240
  in China, 178–79
  Dallas, Texas, 172–73
  De Brouwer incident, 215–16, 220, 225–26, 291
  discussing with others, 181–82, 270, 272
  Eskimo Scouts, 163–65
  in Germany, 180
  Gulf Breeze, Florida, 166–68, 292
  Malmstrom Air Force Base, 168–70, 200, 208, 237, 290
  Mansfield, Ohio, 162–63
  in Mexico, 180, 240
  missile interactions, 168–72, 225, 237–38, 290–91

  in Peru, 179
  Phoenix Lights, 86, 165–66, 210, 225, 253, 292
  sensor systems and, 210–11, 269
  in Soviet Union, 168, 170–72, 173–75, 225–26, 237–38, 240, 291
  Torres incident, 190–92, 193, 195, 198–200, 211, 213, 225–26, 240
  in Turkey, 179–80
  in United Kingdom, 155–60, 189–93
  in Vietnam, 177

UFOs:
  crashes of, 144–45, 222–23; *see also* Roswell incident
  extraterrestrial hypothesis and, 228, 229, 273
  reverse-engineered, *see* reverse-engineered UFOs
  support for scientific investigation of, 270, 293
  technological advances and, 208–10
  triangular, 251–53
  variety of types of, 228, 229, 269
  see also extraterrestrial encounters

United Kingdom, 183–95
  Condign Report in, 192–93
  crop circles and, 185, 193–94
  MoD UFO office shut down in, 194–95
  UFO files released in, 187–88, 190, 195, 240
  UFO sightings in, 155–60, 189–93

unmanned aerial vehicles (UAVs), 209–10

Valdez, Gabe, 263–64

Vallee, Jacques, 17–18, 42, 46, 73, 220, 229

Vandenberg, Hoyt S., 123, 124

Vegas Garcia, Clemente, 180

Vietnam War, 46, 87, 135–36, 177

Von Ludwiger, Illobrand, 180

Waldheim, Kurt, 220

Walker, John, 207

Walters, Ed, 166–67, 168

Warren, Elizabeth, 83

Warshaw, Richard, 30

Watergate scandal, 124–26

Weaver, Richard, 109

Wedel, Janine, 84

Weinbrenner, George, 250

whistle-blowers, 87–89

Whitehurst, Fred, 87

INDEX

White Sands Missile Range (WSMR), 47–48, 106, 243

White World, 80

Widnall, Sheila, 109, 246, 271

Wigner, Eugene, 117

Wood, Bob, 17, 126

Wood, Ryan, 126

Woodward, Bob, 125

Wright-Patterson Air Force Base, 97, 175, 249, 253

Yanacsek, Robert, 162

Zeidman, Jennie, 162

Zigel, F. Yu, 33–34

Zuckert, Gene, 200

CONTACT INFORMATION

Contact with the author can be made through the Web site:

www.JohnBAlexander.com.

Please note the middle initial is important as there are other John Alexanders.

Anyone who has firsthand knowledge of a UFO case, but was sworn to secrecy, or otherwise threatened by bona fide U.S. Government officials, is encouraged to come forward. As indicated in the book, it does appear that some agents overstepped their authority and simply told people not to talk about their experiences. Many people have spoken out, and no known adverse consequences have been taken against anyone. If your claims can be validated, I will assist in obtaining permission for you to openly discuss that information.

A single point of entry for U.S. Government information about UFOs can be found at the Web site www.dod.mil/pubs/foi/ufo.